I0638022

236 POUNDS
OF CLASS VICE
PRESIDENT

ALSO BY JASON MULGREW

Everything Is Wrong with Me

236 POUNDS OF CLASS VICE PRESIDENT

a memoir of
teenage insecurity,
obesity, and virginity

JASON MULGREW

HARPER ● PERENNIAL

NEW YORK • LONDON • TORONTO • SYDNEY • NEW DELHI • AUCKLAND

HARPER ● PERENNIAL

I have changed the names of some individuals, and modified identifying features, including physical descriptions and occupations, of other individuals in order to preserve their anonymity.

All photographs throughout are courtesy of the author unless otherwise stated.

P. S.™ is a trademark of HarperCollins Publishers.

236 POUNDS OF CLASS VICE PRESIDENT. Copyright © 2013 by Jason Mulgrew. All rights reserved. Printed in the United States of America. No part of this book may be used or reproduced in any manner whatsoever without written permission except in the case of brief quotations embodied in critical articles and reviews. For information address HarperCollins Publishers, 10 East 53rd Street, New York, NY 10022.

HarperCollins books may be purchased for educational, business, or sales promotional use. For information please write: Special Markets Department, HarperCollins Publishers, 10 East 53rd Street, New York, NY 10022.

FIRST EDITION

Designed by Michael P. Correy

Library of Congress Cataloging-in-Publication Data is available upon request.

ISBN 978-0-06-208083-7

13 14 15 16 17   OV/RRD   10 9 8 7 6 5 4 3 2 1

for my wife, monk

and to a lesser extent, for the students,
faculty, staff and alumni of sjp

(as you all didn't have to live with me during this process)

contents

Accessary to Bribery   1

Summer of Growth   21

Football Line   47

Master of Crabs   61

Philadelphia's Finest   75

So Let Me Introduce You to the One and Only Billy Shears   91

The Six Most Influential Songs of My Teen Years   99

Lazy Rider   109

The Summer of Magical Drinking   133

The First Six Loves of My Life   155

My Aim Is True   167

236 Pounds of Class Vice President   181

Acknowledgments   203

236 POUNDS
OF CLASS VICE
PRESIDENT

accessary to bribery

It is 1992. It is the Philadelphia City Spelling Bee. I am twelve years old. And I am on fire.

Diaphanous? Sure. *Hors d'oeuvres*? Love 'em. *Proscenium*? Please—where I stood just a few months earlier when I reinvented the role of Laertes in the seventh grade class's production of Hamlet. *Bludgeon*? You mean what I'm currently doing to the competition?

I am the sole representative of my Catholic grammar school, Our Lady of Mount Carmel. Having dispatched all comers in OLMC's bee a few weeks prior, I am now here, in the thick of competition in the city's spelling bee. When we started, there were 130 of us, nerds united by our affinity for video games, comic books, and/or bedwetting. But once the bee began, it was all business. We watched and took secret delight when our fellow geeks, our friends for the day with whom we had spoken fondly about our favorite Pop-Tarts just moments before, were ousted from the competition because of their foolish, incorrect spellings.

And now only a dozen of us remain. Of those, the last seven standing will be invited to the Pennsylvania State Spelling Bee, giving them a shot at the National Spelling Bee. For the junior high geek, it doesn't get much better than this (as any sort of intimate/romantic relationship is still ten to fifteen years away and will require the use of a credit card).

But for one junior high geek, it does get better. You see, for me, there is more at stake than merely a spot in the state spelling bee and a chance to become America's best speller.

For me, there is a Brutus at stake.

For nearly as long as I could remember I had wanted a dog. The allure of having a pet that I could spend time and have fun with, that I could take care of but that would also act as my protector, that I could perhaps teach to do simple tasks like fetching the remote control or pissing on my little brother and his belongings, had landed "dog" at the top of every wish list for birthdays and Christmases year after year during my childhood.

I had had other pets. There was a turtle that lasted two weeks-ish when I was six. Around age eight, there was a snake named Ziggy that had had slightly better luck. And there was my longest-serving pet, a hamster named Boojee (pronounced *BOO-zhee*), who lived for about two years just after Ziggy's passing. All of them fine, sure, something to do, something to play with—but not dogs.

(I had never considered a cat as a potential pet, not only because I was allergic to them but because I never understood the appeal of a cat. Vapid and indifferent, interacting with a cat is not unlike playing with a plant that shits. So, no thanks.)

In seventh grade, just about a year after we lost Boojee, I began taking my interest in getting a dog to another level. I scoured the classified section of the *Philadelphia Daily News* every day, reading ads for different breeds, trying to decide what kind of dog I'd get. Pussy dogs were out. There would be no poodles or Pomeranians or Yorkies or the like. The goal was not to cement myself as the Biggest Gaybird on Second Street, my South Philadelphia neighborhood (which, counterintuitively, was not just one street but a collection of streets around a main thoroughfare). Rather, because *I* was such a pussy, I needed a tough dog to both enhance my image ("Hey, look at the kid with that mean dog! He must be a mean dude, obviously!") and serve as protection ("Hey, look at the kid with that mean dog! I'll have to wait until he's without that dog to kick his ass, obviously!").

But at the same time, my dog couldn't be too intimidating. While it was fine for bad guys to fear my dog, I didn't want my friends and family to be afraid of it. My buddy Larry had a Doberman called D'Ogee (Get it? *d-o-g*?), and every time my friends and I were at his house, our plan was simple: don't get your face eaten off by D'Ogee. We'd sit statue still, eyes fixed on D'Ogee, barely moving, barely breathing, lest D'Ogee take any movement as a sign of aggression and subsequently clamp his jaws around the nearest friend's genitals while clawing and head-butting the rest of us. I feared hell less than I feared D'Ogee, and I didn't want any friend or visitor to my home to feel that deep, dreadful terror in the presence of me and my dog. Although image enhancement and protection were important, I was looking for a buddy, a pal, a comrade-in-arms. My dog and I were going to be a one-two punch, known to all in the neighborhood and as popular as other duos— like Laurel and Hardy, or Cheech and Chong, or, I don't know, Starsky and Hutch.*

I was looking for tough, but fun, approachable. Therefore, in addition to knocking poodles/Pomeranians/Yorkies off the list, Dobermans, rottweilers, German shepherds, pit bulls, and any other dog part wolf or werewolf was out. But neither would golden retrievers and labs fit the bill because, though large, they lacked toughness. Bulldogs were too wrinkly. Great Danes were too big. Collies (Lassie? Really?), basset hounds (silly looking things), Chihuahuas (anything I could throw farther than a football was a no), Dalmatians (like the fucking Disney cartoon?)—I crossed out each as I combed through the classifieds.

But then, there, after many days of weighing the pros and cons of each breed, there it was, right in front of my face, right there near the top of the ads: the boxer. A lean, fit, muscular dog, the boxer certainly looked tough and scary, but not so tough and scary that friends would fear him even while he slept. And based on my subsequent research, the boxer could also be playful, boisterous, and good with kids. Those adjectives would never be used to describe me (aside from maybe "good with kids," on account of all my little cousins). My boxer and I would complement each other. We'd make the perfect pair, just like . . . well, forget it. I'm out of duos.

The matter was settled. I wanted a boxer. And I wanted to name him Brutus. I'm not sure where the name came from or what was its significance, but it just sounded right, what with

---

* Apparently, I am only familiar with popular duos from before 1980.

the alliteration and the root *brute* and all. If I was going to have a badass (but cool) dog, it needed a badass (but cool) name.

The attainment of Brutus thus became my singular obsession for much of 1991 and 1992. I learned as much as possible about boxers. I saved as much money as I could. I traveled to pet stores to see if they had boxers or to peruse the dog accessories that I would one day purchase for Brutus.* Indeed, in my world, it was all Brutus, all the time.

Where did this doggie desire (not what it sounds like) come from? Let us consider.

The author (left), enjoying some time with a few of
his cousins and brother Dennis (right).

---

* It never occurred to me to visit a pound and get a rescue dog. I don't know if this is because of a rebranding campaign by the ASPCA over the last decade, but it used to be that the pound was not a place you went in order to save a dog's life after being inspired/guilt-tripped by a Sarah McLachlan commercial. Back then, the pound was dog prison, where bad dogs who couldn't behave went to . . . well, you know.

Was I lonely as a child? No, if only because that simply wasn't possible. Though I only had two siblings—my younger brother Dennis and my little sister Megan—I grew up in a large, extended Irish Catholic family. My dad was one of ten kids and my mom one of six, so I had a shit-ton of aunts and uncles and (literally) dozens of cousins. Every weekend there was some large family function that required my attendance (as well as that of Dennis, Megan, and our mom and dad), functions at which the men drank beer and smoked cigarettes, the women gossiped and appraised the potato salad, and the children ran around like maniacs, flying high on sugar and reined in only by a well-timed "Hey, knock it off!" from Uncle Joey, who would be on his ninth Miller High Life. I was the second-oldest cousin on both sides of the family, and at any given party there were twenty or more cousins ranging in age from thirteen to zero. It was pandemonium.

Since this is Second Street we're talking about, all of these relatives were within walking distance of my home. Second Street—or "Two Street," as it is sometimes called by outsiders—is:

- predominantly Irish Catholic (while Rocky did much for the city of Philadelphia, particularly South Philly, it's not just full of sons and daughters of It'ly);
- blue collar (the most common occupations for adult males on Second Street are longshoreman, roofer, carpenter, electrician, and what I will generously call libertine, by which I mean alcoholic, gambler, petty criminal, fighter, junkie, or some combination thereof);

- and, perhaps more than anything else, provincial.

On Second Street, the apple never fell far from the tree. Hell, sometimes the apple stayed *in* the tree. My grandparents on my father's side had one child living with them and three children and their families living on their very street. If you're keeping score, that means four of their ten kids moved maybe five hundred feet from the house in which they grew up. My grandparents on my mother's side had two children living with them and two children and their families just one block away. So four of their six kids could be reached not only by telephone, but via a good, from-the-diaphragm yell out the window.

Everyone knew everyone else in my neighborhood. If you were born into Second Street, you were automatically plugged into a network of hundreds, spanning generations. You've heard of six degrees of separation, right? Where I grew up, we never got past two degrees. If you were to eavesdrop at a local bar, you might hear something like:

MALE 1
You know Billy McGowan?

MALE 2
Nah, I don't think so.

MALE 1
Yeah, you do. He's the one from Third and Mifflin with the lazy eye that got in a fight with Coyote from Sixth and Porter?

MALE 2

Oh, yeah, I know him. His cousin is Franny Boyle, right?
Good pool player.

So . . . no, the reason that I so desperately wanted a dog was
not because I was lonely or in dire need of someone to hang out.
Loneliness, solitude, peace and quiet: these things would have
been welcomed.

Was I unloved as a child? Nah. Not that, either. I must con-
fess, however, that when I was eight, my parents began a two-
year-long divorce process during which my mom, brother,
sister, and I moved into my grandmother's house, where my two
uncles also lived. But stop before you jump to conclusions and
say to yourself, "Aw, poor little guy. He wanted a little doggie because he came from a broken home." That's not really the case.
The divorce was not without its bummer moments, but I still
saw my dad and his family regularly, and throughout the split
both my parents remained present and loving and all that. Furthermore, all of my relationships as a kid—with family, friends,
teachers—were appropriately fulfilling and appropriately appropriate. Sorry, but this is not that kind of memoir.

Maybe I was just bored? Yes, I think we have a winner.

Don't get me wrong; I had shit going on. I had a job delivering papers, and in the summer, in addition to the paper
route, I worked part-time at a camp. I wish I could tell you that
I spent my time as a counselor smoking cigarettes and fingering
girl counselors, but this "camp" was held at my former elementary school, smack in the middle of the city. It was a 9 a.m.–to–
1 p.m. gig during which I made sure that kids didn't run too

fast in the schoolyard. And then I drank a lot of chocolate milk during arts and crafts time, which was held in the basement of the rectory. So, a pretty shitty camp, really.

I had non-work interests, too. Certainly, I enjoyed comic books and video games, as well as eating—and if you had seen me at this time, you could have guessed that these three were chief among my hobbies. Seventh grade was shortly before I discovered masturbation, which I also took a liking to (but we'll get to that later).

I was crazy about music. I played saxophone for a bit, but lessons were expensive, and after two years I could only play "Happy Birthday." So I dropped the sax. For a fat, nerdy white kid, my tastes in music were surprisingly urban. I listened mostly to rap, hip-hop, and R & B. I especially enjoyed *The Quiet Storm*, a slow jams program on Power 95 FM, which I was 95 percent sure was what black people listened to while they had sex. Later, I would discover (and fall in love with) rock 'n' roll.

I liked being funny and funny things. I know this is not a real hobby,* but it's the truth. I was an extrovert, a class clown who learned early in life that I could make people like me and get some positive attention by making fun of myself. So that was my shtick: Hey, everyone, look at me! I'm chubby! I'm a nerd! Isn't this hilarious? (Trust me, it was funnier than this. Or maybe you just had to be there.)

Finally, I really, really liked sports. Not so much playing sports, but rather watching the games and studying the box scores. I played two years in Little League, and looking back I'm comfortable using the word *nightmare* to describe that experi-

---

* Nor is it a very good sentence.

ence. I cried more times than I had contact with the ball, and, though I'm no scout, I'm pretty sure this is not a good thing. I also tried out for the local kids' football team, but because I was bigger, and the teams were based on weight and not age, I would have had to play with kids two and three years older than I was. After one informational session, my dad and I never went back. I had no desire to get my ass kicked *on* the football field by the same older kids who wanted to kick my ass *off* the football field.

I don't mean to sound like a twerp, but I really didn't have to study for school, which, for the most part, I found pretty boring. Once a week, I left Our Lady of Mount Carmel and attended a program called "MG," which stood for "mentally gifted." In today's more politically correct environment, I'd guess this program is now called "Learning for Learners!" or "Yay for You for Enjoying School!" For MG, I went to the local public school, a terrifying monolith of a building that looked more like a Communist-era Eastern European administrative center than an elementary school, and hung out with other geeks from different schools in the area. There, we worked on "out-of-the-box" projects, like a game in which we picked stocks and tracked our returns, and a longer-term assignment in which we were each given a Senate seat and argued a proposed bill (I was the esteemed senator from the great state of Montana: "Mountains and Trees: Boy, Do We Got 'Em"). All this was done under the direction of the two hippie teachers, Ms. Blumenthal and Mr. Hansen, who undoubtedly spent their free time smoking some serious pot and painstakingly dissecting *Blonde on Blonde* line by line. There was little structure. We were encouraged to do whatever we liked: read a book from the in-classroom library, do our homework, discuss current events with each other and the teachers. I chose to spend my free time playing hours upon hours of Oregon Trail on our car seat–size computers and talking to the black kids—the first black people I had ever spoken to—about music and all things Eagles, Phillies, and Sixers—but not so much Flyers. (One of the first things I learned about black people from firsthand experience: they're not really into hockey. Which was fine, because neither was I.)

The future senator from Montana.

And yet still, I was, frankly, bored to shit.

There are two things you must understand, dear reader. The first is that this was the early nineties. Whereas twelve-year-olds nowadays are busy with soccer matches, SAT practice courses, Mandarin IV, flying lessons, training for a half-marathon, volunteering at the senior center, practicing for the upcoming piano/harp/xylophone recital, and leading various seminars on time management, twelve-year-olds in 1992 didn't do shit. Sure, we went to school, maybe played a sport, maybe had a little job, but we weren't twelve years in training to become the next president of the world (yes, the whole goddamned world). I'm sure that if I came of age now, I'd have been diagnosed with ADHD at age three and would have been so shot up with Adderall over the course of my lifetime that I'd be writing this on the roof of my building right now, taking breaks only to scream that I'm the second son of the Virgin Mary and to see how quickly I could do one hundred jumping jacks. After school, work, and sports, there wasn't much for your average kid of lower-middle-class origins to do back then.

The second is that twelve is a strange age for a kid—particularly for a boy. No longer a child, not yet a teen, you're just hanging out, waiting for puberty to change the whole motherfucking game. Until then, there's a weird void. Perhaps if I had discovered masturbation a year earlier, I wouldn't have focused so much energy on getting a pet dog. Alas, I can't complain about how *that* all worked out.

This confluence of events and circumstances—my boredom, my age, my precociousness—made my desire to get me a Brutus overwhelming. For years, I had wanted a dog. But now, I was serious. There was only one teeny-tiny problem: my mom would have none of it.

My mom, Kathy, was not an unreasonable person. She was a good woman, a tough broad who stayed with my dad during his hard-partying days until she'd had enough. She was now raising three kids by herself. She did clerical work at the UPS office by the airport and enjoyed oldies and the occasional chocolate marshmallow ice cream (with her Philly accent, pronounced MOSH-*mal-lah*; see also *library*, as in LI-*behr-ree*). I loved and respected her and tried my best to be a good kid, to do well in school, to help around the house, to not be a total monster to my brother and sister. I also got a job so I could have some pocket money. I'm not saying I was the perfect child, but I was pretty close.

Her objections to Brutus ownership were not without merit. They fell into three camps: dogs are a lot of work, dogs cost a lot of money, and we live in a small house without a proper yard. Well . . .

Yes, dogs are a lot of work. But, Mom (this is me talking to her), was I one to shy away from work or extra responsibility? As one who has witnessed my obsession on a daily basis, do you think I'm going to get a dog, lose interest immediately, and move on to something else? Has any pet or other living thing met its premature end under my watch or at my hands? Except for that turtle, like, six years ago? (Which Dad found while on a crabbing trip, might I add, so it wasn't meant to be in a fish tank and most definitely went to a better place.)

Yes, dogs cost a lot of money. But, Mom, not *that* much money! I have a little bit of money from the paper route, which *of course* I would use for Brutus! And next year, when I graduate, I'll get more cash from my graduation and the party for it, which *of course* would go into the Brutus fund!

Yes, we live in a small house without a proper yard. But . . .

. . . She kind of had me with this one. We lived in a row house with a "yard" that was no more than an eight-foot-by-three-foot slab of concrete shared with our next-door neighbors, an elderly couple we estimated were in their 130s. It was a place where we put garbage until trash day came around and where either my brother or I would pee if we really had to go and the bathroom was occupied. No, this was not ideal for a dog that could grow to seventy pounds.

But Mom (now we're switching back to me talking to her), it can be done! We know other people who have even larger dogs—and the same size yard! And I would make sure Brutus gets plenty of exercise! And maybe that would help get me in better shape! See? Everyone wins!

And yet despite my best efforts, she would not budge. But I was not deterred. I knew this would be a long fight, and I settled in for a siege. However, my campaign would be shorter than I had expected. Because I had just won the Our Lady of Mount Carmel Spelling Bee.

Academic excellence was very important to my mom. Or at least, the *pursuit* of academic excellence by her children was very important to my mom. Both my parents were smart, but not exactly learned. Both had graduated from high school. And both read the paper every day. But aside from that, I don't recall my mom reading anything that wasn't a *Star* (or *Star*-type) magazine, or my dad reading anything other than a transmission manual or *Popular Mechanics*. Aside from the comic books that Dennis and I owned and a set of encyclopedias that may have been released before the end of the Vietnam War (and was only about 70 percent complete), there were no books in our house.

But neither my parents' lack of a degree nor the near-complete absence of reading materials in our home meant that Dennis, Megan, and I couldn't be learned. We were not only going to go to college, but we were going to go to good colleges. And in order to do so, we had to be academic superstars. Therefore, the need to do well in school was instilled in us from an early age. This was never forced—it's not like my mom locked us in our rooms or beat us with hangers if we botched our recitation of the multiplication tables—but she did bribe the shit out of us.

I don't think that a parent bribing a child is all that uncommon. I don't have any children myself, and whenever my friends have kids I sort of drop them from my life because, Jesus Christ, all they want to talk about is their kid walking or talking or shitting, and it's the most boring thing in the whole entire world to listen to. But I would guess that bribery is one of the basic and most effective tools of parenting. Clean your room and you can have extra dessert. Do the dishes all week and you can stay out late on Friday night. Do well on your final report card and maybe you'll get that new bike, just in time for summer.

My mom bribed us with the usual perks: food, toys, money, extended TV viewing, allowing friends to sleep over. And it worked. I was doing well in school. Dennis entered the MG program in third grade (whereas I didn't get in until fifth!). The jury was still out on Megan, who was only in first grade. While it's not like our mother's bribes had everything to do with how we approached and succeeded in school, there's no doubt that her chocolate chip cake was a powerful motivator. We operated under the unspoken arrangement that doing well in school would be rewarded: get good grades, get good shit.

This is why I didn't tell my mom about the Our Lady of Mount Carmel Spelling Bee. It didn't affect my grades, and I didn't think it was a big deal. Because I had participated the year before but lost—the bee included the entire junior high, so an eighth-grader won that year—I didn't want to set myself up for failure by talking a big game and not delivering. It wasn't until after the bee was over and I'd won first place that I casually mentioned it to my mom at the dinner table.

"Really? That's great! That's really, really great, Jase." She then fired off a number of questions: Who else was in it? (I don't know, maybe twenty-five kids?) How many rounds did it go? (I don't know.) What word did I win on? (I forget.) Did I get a trophy? (Nah, just a piece of paper.) Why didn't I tell her about it before? (I didn't think it was a big deal.)

But it was when I explained to her what came next—the Philadelphia City Spelling Bee in a few weeks, from which a handful of finalists would advance to the Pennsylvania State Bee, the winner of which would qualify for the National Spelling Bee—that she got really excited. Her child, potentially the best speller in the whole city or the whole state or even (dramatic gulp) the whole country? Jesus, Mary, and St. Joseph. This was big, big news. And it called for a big, big bribe.

My mom and I made a deal. If I reached the state spelling bee, I got Brutus. This meant that I did not have to win the city spelling bee. I simply had to be among the kids who went on to the state bee. Not even win, just shoot for being among the top seven? Yeah, I could handle that.

I took the spiral-bound book of spelling-bee words that I

was given after the school bee and went off to my room. It was time to get serious.

"Panegyric. P-A-N-E-G-Y-R-I-C. Panegyric."

Panegyric? What the hell is that? I've never seen or heard that word in my life, and I'm about 50 percent sure it's made up. She may have spent most of the competition in her seat sucking her thumb, but damn, that girl is a good speller.

It's getting down to the wire. We're on hour two of the bee and only twelve are still standing. Being in the top seven was all I needed for Brutus.

It's getting warm in the auditorium. Parents, legal guardians, and loved ones are seated in the audience. The pressure is mounting. And the words are getting harder. At first, everyone got nice, easy words, perhaps to relax the participants. But then it was time to separate the wheat from the chaff. In those middle rounds, people started dropping out, losing either on difficult words or crumbling under pressure and blowing the easy ones. I was able to get a good rhythm going, and though I got some doozies (damn you, French, and your stupid *maître d'*), I was buzzing along at a pretty good clip. I didn't know if I could win, but top seven? Yeah, I could do that. Especially now that it was so close. I could spell all day, baby.

After the thumb-sucker, it's two more and then me. A kid who looks quite a bit like me, except with a catastrophic lazy eye, steps up to the microphone.

"Michael, your word is *deleterious*."

That's a good one. Not too easy, but not too hard—it's just

delete with *–rious* on the end. And Michael nails it. Good for him. I guess.

Next, an Asian girl. She gets *recidivist*. I think I could probably spell this, but you never know until you get up there. It's one thing to spell the word in your head at your seat and another to stand in front of the judges and all the parents and do it aloud. I don't usually have trouble with consonants—I know a *C* from an *S* and a *K* from a *CH* or *X*. It's the vowels that sometimes trip me up. *E* vs. *I*, *A* vs. *O*. These were the spelling traps that kept me up at night.

The Asian girl spells *recidivist* right. Now it's my turn.

"Jason, your word is *accessory*."

There are audible groans from the other participants. This has happened before: occasionally, a softball sneaks in during the later rounds. And *accessory* is definitely a softball. I got this.

But . . .

But there was something strange in the way the judge pronounced the word. See, in Philly-ese, *accessory* is a hard, ugly word: *ak-sess-uh-ree*. The judge, who has been devoid of any accent throughout the bee, made it sound softer, a bit more flowing: *ah-sess-ah-ree*. Sort of like she was combining *ass*, *ess*, and the end of *rotary*. *Ah-sess-ah-ree*.

I know it starts *acc–*: that ain't gonna trip me up. But is it *–ory* or *–ary*? I thought it was an *o*, but she made it sound like an *a*. Is *accessory* a different word from *accessary*? Is she using the latter? Like, maybe one is a clothing *accessory*, but the other is an *accessary* to a crime?

Alright, deep breaths. Focus. It's either *o* or *a*, and I have to pick one. It's gotta be *o*, right? But maybe not; *–ary* is a common suffix, too: *necessary, arbitrary, literary*. All I know is that when I say the word, the first syllable (*ak*) and third syllable (*uh*) are not alike. The way the spelling bee emcee says it, the first (*ah*) and third (*ah*) syllables sound identical. And since I know the word starts with *a*, that's what I'm going with.

"Accessary. A-C-C-E-S-S-A-R-Y. Accessary."

"I'm sorry, Jason . . ."

I was devastated. It was not just the loss, but the word. On the car ride home, the next day at school, several days later, I kept turning it over in my mind. *Accessory*? Really? I'm in M-fucking-G, but I can't spell *accessory*? Not something like *antediluvian* or *chartreuse* or some other word that I could be proud of losing on? That I could look back upon with respect as a formidable foe that truly and fairly bested me? Losing a city spelling bee on the word *accessory* is like going off to fight the Nazis in 1944 but coming home a week later because a bad crêpe you had in Paris gave you the runs.

There is no shame in losing, I was told. You did a fine job, I was told. We are all very proud of you, I was told.

We'll get you Brutus next week: this I was never told. Like I said, my mom was a tough broad.

summer of growth

The loss of Brutus was difficult, but the summer between seventh and eighth grades saw new developments that allowed me to move on. The most important related to my heightened awareness of my genitals and how they worked.

I had heard all about masturbation, but most of the hearsay revolved around the abstract *concept* of masturbation. That is, there was much talk of guys jerking off, wanting to get jerked off, stop being a jerk off, and so forth. In fact, *jerk off* was among the top twenty phrases I used or heard on a daily basis. Even the adults I knew used *jerk off* to describe a guy who was an asshole (though they generally preferred the more old-school *strapper*, which I have never heard used even once outside the adults in my family and neighborhood).

But what jerking off actually involved was unknown to me. While I had a rough idea of what it was, I had a number of questions about the physical process. I knew that touching one's bird felt good.* And I knew that things touching one's bird also felt good. (Two years earlier, while vacationing on the Jersey Shore, I spent an entire week in the exact same spot in the pool after discovering that the water jet shooting on my bird felt just oh so marvelous.) But when jerking off, do you just pull on your bird? I mean, do you literally jerk it? Doesn't that hurt? I could see gently rubbing it, but tugging on it? Really? Even bonerized, it seems like a delicate thing to be yanking on.

And this jizz stuff . . . are we sure it's not pee? I've heard that it's supposed to be white, and pee is yellow, but I've peed not-as-yellow after drinking a whole bunch of water. Is that jizz? I was pretty sure that jizz and light-colored pee were different things. But you never know.

And after jerking it for a while, then the jizz comes out, right? That's the endgame here? How do you know when or if that's going to happen? What if it doesn't? How long should I keep going? How do you know when to stop? When your arm gets tired?

And does the jizz come, like, flying out? What kind of speed are we talking here? Faster than the fastest pee? So fast that I'd feel a kickback in my hips, like after shooting a gun? Or does it just dribble out, like those last drops after a long pee? Do I have to wring out my bird to make sure it's all out of there? What if there's some left over?

And does it shoot all over? What kind of damage radius are

---

* We called the penis "bird" in my family. I think this is a Philly thing, but if you grew up outside of Philly and called your dick your bird growing up, please email me at jason@jasonmulgrew.com. Perhaps we can get together for a beer.

we talking? You know how a bottle of household cleaner has the "stream" or the "spread" option? Which is it closer to?

The whole thing was confusing and intimidating.

Then one early summer evening, my buddies Phil, Vic, and I were hanging out on Phil's front steps when the topic of jerking off came up. Phil and Vic were two of my closest friends. Phil was athletic and cocky, which meant that most girls liked him and most guys secretly admired and openly despised him. Vic was a bear of a kid, but very soft-spoken, and he possessed an intensely dry sense of humor. Both Phil and Vic knew a lot more shit than I did because they had older siblings (a sister for Phil and a brother for Vic) and access to their older siblings' friends.

I couldn't ask the specific questions that I had about jerking off. To let on that I hadn't been jerking off for years, that I wasn't masturbating several times a day, that I hadn't masturbated three times that day already (let alone that I had never done it even once) would have been a grave error and significantly detrimental to what little social standing I had. So as Phil and Vic talked about how awesome jerking off was, I sat back and said things like, "I know, right?" and "totally" and "oh, that's the best." But despite my bluffing, I did manage to pick up one little nugget. The winning technique wasn't to jerk your bird, but to slide your hand up and down around it. "I think of it like you're fucking the tube of a toilet paper roll," Phil said at one point, miming the action. "But instead of humping it, you let your arm do all the work." Well, there we go.

Inspired, I bid my friends farewell at the first opportunity and raced home to see what all the fuss was about.

First I had to decide where to jerk off. As everyone was at home, my options were limited. I considered the basement, but it was unfinished, grimy, and, frankly, scary. My bedroom was another option, but there was no lock on the door, and I was often disturbed while in there. I thought briefly about the roof ("Oh look, Frank—there's the oldest Mulgrew child masturbating on the roof"), but knew there was only one viable option: the bathroom.

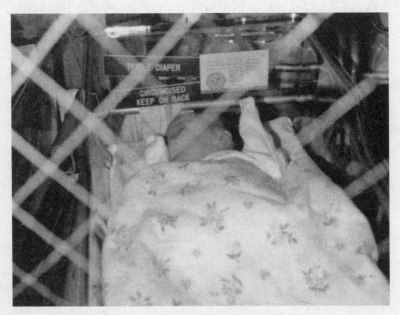

I still have to sleep on my back. I think they
might have screwed something up.

The bathroom made the most sense. If I was going to murder someone, it would be in the bathroom. It was the easiest room to clean: everything was tile. There were ample amounts of both cleaning supplies and toilet paper. I could run the shower to drown out any noise. I could flush almost anything down the toilet. True, it was the only bathroom in our house and was therefore highly trafficked. But the lock was good and strong, and unlike my bedroom—into which my mom, brother, or sister had no problem barging at any time—there was no interrupting someone in the bathroom except for a top-flight emergency. If the bathroom was the best room in the house to murder someone, it was the best room in the house in which to jerk off.

I sat on the toilet. Bonerization was no problem; strong breezes, sudden changes in barometric pressure, and 99.97 percent of my waking thoughts were enough to give me an erection. I formed the toilet paper tube with my hand and began the sliding process. At first I was concerned with the end product. What would happen when or if the jizz came out? But then a funny thing happened: during the course of the (let us call it) *sliding*, I grew less and less concerned about the final result. Or about my homework. Or about where I was, what time it was, whether there was a God, and whether He was watching me disapprovingly. I was rather focused on the, uh, process.

And so I kept going. I kept going. And kept going. And then: boom. Or shall we say:

BOOOOOMMMMM!!!

[Makes slight choking noise, blacks out for a few seconds.]

Now, I'm not going to tell you how wonderful it is to have an orgasm, because a) I'm not a good enough of a writer to adequately describe it, and b) you're not a fucking idiot. I will tell you that my life was henceforth divided into two distinct eras. There was BC and AD for Western civilization; there was BJ (before jizzing) and AJ (after jizzing, or, *anno jizzino*, if you prefer) in my own world. Not quite the birth of Jesus Christ to separate the two, but pretty damn close. It was the most significant of my critical life moments to that point, ranking above the time I learned there was no Santa and finding out that the really friendly ice cream truck guy who always complimented me on my hair was actually Class 2 felony ice cream truck guy. I knew that from that point forward, nothing would be the same. Things were different now, and there was no going back.

I took so much to my new hobby that I am certain there were bordellos in the Old West during the height of the Gold Rush that didn't see as many orgasms as the cold, blue tile floor of that bathroom on Third Street in South Philadelphia during the summer of 1992. Not only were all the questions that I'd previously had about jerking off answered immediately, but I went from novice to expert in no time, experimenting with a number of different approaches and tactics, including going southpaw (masturbating with the left hand), doing the invert (flipping your hand over so that you're beating off with your palm on top as opposed to the more standard palm-underneath technique), meeting the Stranger (sitting on your hand until it fell asleep and then masturbating, creating the impression that a stranger's hand is jerking you off), and a personal favorite, meeting the Jersey Stranger (while standing in the shower, sticking your arm out and wrapping it around the shower curtain, as though a stranger, presumably a stranger from New Jersey, is reaching into the shower to jerk you off).*

Thankfully, at that time, I knew nothing of the "grundel button," the area between the scrotum and butt hole also known as the "gooch," "chode/choda," or "t'aint" ("t'aint your balls,

---

* You could also combine the techniques as you saw appropriate—e.g., a lefty-invert or an invert-stranger. The lefty-invert-Jersey Stranger was just too complicated, however: way too many moving parts.

t'aint your ass"). Pressure on this area immediately prior to ejaculation—"pushing the grundel button"—results in an orgasm powerful enough to tilt the earth on its axis and has caused certain people to spontaneously combust. This is advanced shit: AP Jerking Off. Had I learned of the grundel button just as I started beating off, I would have quit school and run away to dedicate myself solely to masturbation, living a contemplative life somewhere among lush, verdant hills, possessing or desiring nothing of this earth, save for a lot (a whole lot) of moisturizer.*

Practicing for later in life, perhaps? Also, wearing the dumbest Halloween costume ever. I guess for Halloween that year I was a penis in sweatpants with a stick between its legs?

---

* A word about moisturizers and other masturbatory lubricants: if it says "for external use only," put that shit down immediately. I still get tingles in my dick every few months because of a beat-off session back in '96 during which I was a little too generous with some SPF 70, some of which was lost down the rabbit hole. So stick to hand cream or Vaseline, and be sure to read every word of the label of anything you plan to slather on your penis. You're welcome for the best advice you received today.

Adding gasoline to the fire of my masturbatory mania was "the chip," for the summer that I discovered jerking off was the same summer that my family decided to start stealing cable TV (kismet!). A household did this by getting the chip installed in its cable box, thereby affording access to all the premium movie channels, pay-per-view boxing and wrestling events—and all the porno channels—for free. I could now watch, twenty-four hours a day and from the comfort of my living room, strangers having sex on television.

I had no personal experience with sex whatsoever. In sixth grade, I had had a girlfriend named Kara. We kissed a few times, and one time, she kissed my neck. Which was awesome. But then Kara broke up with me, and I had been on a major dry spell since. Other friends may have been feeling boobs (*on top* of shirts), but I was still hoping to kiss girl number two.

My knowledge of sex was more pathetic. In terms of the female anatomy, I'd seen a *Playboy* or two in my day, so I was familiar with the setup: girls had boobies, heinies, pubes. Got it. Though the nudie mags did inspire something in my pants, I was more fascinated by the female body than turned on by it, viewing it scientifically ("Hmm . . . very interesting, the complete lack of bird and balls that we have here . . .") rather than lustily ("I would like to do things to that, though I'm not sure what type of things").

In keeping with centuries of Irish Catholic tradition, I didn't speak to my parents about sex—not once, not ever. My dad was a tough guy, a longshoreman who worked in sweltering heat and frigid winds and who had seen friends and coworkers die in horrible accidents on the pier. But asking him to tell me all about how babies are made might make him faint. Asking my mom about sex was out of the question; to do so would result in being shunned by the entire community. I'd be taken from my room in the middle of the night, forced into a van, and dropped off on the side of a mountain with some beef jerky, holy water, and a rosary, never to be spoken of again.

School didn't offer any insight either. Once our class was split into boys and girls. The girls were taken into a room to watch a video about periods. The guys were talked to by Mr. Bruno, who taught eighth grade and was one of the few male teachers at the school. He said that we should be nicer to the girls because we were becoming young men and that part of being young men meant being nicer to young women.

Most of what I knew about sex came from the schoolyard, secondhand knowledge passed on from what so-and-so learned from an older sibling or friend or while away at camp. Even so, fingering or getting a hand job was the ultimate sex act. Actual fucking was the furthest thing from our minds.

But the porno channels and the sex they showed changed all that. Actual fucking went from the back to the very front of my mind. If I had any doubts about how the process of sex worked, they were cleared up after only a few minutes of viewing the Spice Channel.

See, in our society, we take pornography for granted. But for hundreds of thousands of years,* the only ways to experience sex were to have it or to think about having it. Now we can see people having it anytime we want. No, better. Now, I can

---

* Maybe even millions: not a big science guy.

find a video of two black chicks having sex with an Asian guy dressed up like a dinosaur in under twenty seconds—and view it on my smart phone while sitting in an airplane bathroom (and no, I don't know this from experience, especially if you were on Delta flight 211 from JFK to Denver on April 6, 2011). For today's male, looking at and masturbating to porn is as unexceptional an occurrence as going to the ATM or downloading a new album. As a matter of fact, I'm looking at two porn videos right now as I type this. They are terrific.

But on that day those many years ago, when I had just started my career as a chronic masturbator and had never before seen actual sex, the effect was startling. This . . . this is sex? This is lovemaking? This is how babies are made? These people looked angry, intent on destroying each other (why is he smacking her in the face with his bird?). All the screaming, all the cursing (now why is she slapping *him* in the face?), all the hair pulling, all the sweating (did she just spit on him?), all the colliding of tanned, fit bodies, all the . . . all the . . .

I thought that sex was taking off all your clothes, getting under the blankets, kissing, and doing some rolling around. Not, for example, sitting in a hospital bed and when the nurse comes to check your temperature she—out of nowhere, apropos of *nothing*—starts giving you a hand job and then a doctor walks in and the nurse starts giving *him* a hand job (Two! Two simultaneous hand jobs! She must be really strong!), and then the three descend into an orgy of madness and yelling and aggression and basically have sex like lunatics. It was appalling. For about forty seconds. And then it was awesome. And then it was really, really awesome.

And then that was it. I was all in. Hooked, hard, and under the spell of the porno channel. I had to be stealthy: I couldn't just plop down on the couch, turn on porn, pull out my penis, and go to town anytime I wanted. But for the rest of that summer, anytime I was alone in that house, channel 35, 77, or 78 was on and my bird was in my hand. Otherwise, I was in the bathroom. And I could have done nothing to stop this beat-fest. (Admittedly, I didn't try.)

In the span of one summer, I went from a sexual neophyte who thought pregnancy was caused by peeing on a girl's butt to a full-fledged, world-class porn monger and masturbator.

Growth, yes.

But the summer was not all jerking off twenty-four hours a day. I was going to summer school.

The Pre-Eighth-Grade Program at Saint Joseph's Preparatory School (or "the Prep," as it was known) was a seven-week course designed to prepare soon-to-be eighth-graders for the rigors of high school through a curriculum of English, math, and, for some reason, gym.

The Prep was King Shit of Philadelphia high schools. It was a hundred-plus-year-old Jesuit institution that produced young men who would go on to become intellectuals, professionals, and leaders in their fields. Graduates of the Prep went to the finest colleges and universities in America, and Prep alumni dominated Philadelphia politics, business, medicine, and law. They also had a crew team (rowing—how sophisticated!) and required all students to take three years of Latin *plus* three years of a modern foreign language. These were facts that I learned from the Prep's glossy admissions brochure, whose cover photograph was of that year's student council officers, handsome young men, smiling brightly, who looked like they could be senators. I'd requested the brochure a few months earlier and had since read it cover to cover eight hundred times, imagining myself hanging out with the student council officers, talking about fancy-pants stuff like art and literature and scotch and Europe. To my uncouth little Second Street self, though I hadn't even seen it in person, the Prep exuded success, refinement, class, and intelligence. And as someone who considered himself the smartest person he knew, I wanted in.

(And "The Prep." Its name alone represented peerlessness. And, let's just admit it, douche-baggery. As much as I was enamored with the school, you can't get away with calling yourself "the" *anything* without sounding like an elitist asshole. Did students at *the* Prep also eat *the* Food in *the* Cafeteria? I hoped to find out.)

It was not all perfect, however: the Prep was exorbitantly expensive. Not on the level of college tuition, but well beyond what my mom and dad could afford. Yet the Prep maintained that though its tuition was high, Mother Prep opened her arms to *all* the best and brightest young men in the Philadelphia region, never turning away a student who possessed the aptitude but lacked the funds: "Where children of bus drivers study alongside children of doctors" was the line they used. (This would piss me off if I were a bus driver. *That* was the "poor person" job they came up with? I understand *janitor* is overused, but what about *dishwashers*, or *carnies*, or even *the perennially unemployed*?) True to its humble roots—it had been founded by a group of Jesuit priests in the 1850s—the Prep remained in North Philadelphia, which, over the past few decades, had become one of the worst, most crime-ridden neighborhoods in Philly. No ivy-covered walls or ivory tower here, folks! We're in the shit! We still got that common touch!

I enrolled in the Pre-Eighth program as a sort of trial. If I loved the Prep, I could then seriously consider attending. If not, I could pursue other options. Those other options, however, were not so appealing.

First, there was public school. There were a number of public high schools in and around my neighborhood, but the estimated time I'd last in any of them before I was murdered in a massive wedgie incident or a lunch money robbery gone wrong was between two and five weeks. The public schools where I grew up were not so much high schools as they were pre-prisons; whereas most students got lessons in Shakespeare or geometry in high school, I imagined that students at the local public high schools took classes like "Hooch: How to Make It and How to Make It Not Kill You," and "The History of Shivs III." These were rough places, and they would eat me alive.

The other option was Neumann, a Catholic high school for boys that was the alma mater of my father, most of my uncles (the ones that had graduated from high school), and my grandfather. If I were to attend, I would be the first of the third generation of Mulgrews to go to Neumann. Furthermore, this was where the great majority of my friends were planning to go, which meant my estimated time of attendance prior to wedgie death was probably more like a year—a year and a half, maybe. And while Neumann cost a little bit of money, the tuition was manageable, and I was almost sure to get financial aid or a scholarship, thus lessening the burden on my parents.

I considered Neumann a back-up plan. A decent back-up plan, but I wanted a piece of the Prep. I was pretty sure I did, at least. That's what Pre-Eighth was for.

It was only a short time before I had fallen in love with, pledged my allegiance to, decided I would do anything to attend, and possibly even fight and die for the Prep. I began to feel this way even before the bus reached the parking lot. While not one to be smitten by the architecture of houses of worship, the Church of the Gesu, rising above the neighborhood's rundown houses, was a beacon of hope that called to all nerdy, socially awkward tweens everywhere: "Come to us. You and your in-depth knowledge of both the Marvel and DC comics universes are welcome here. And please ignore the junkie beating the hooker with what appears to be the business end of the Club. It is safe here. Promise!"

To enter the high school was a powerful, transformative experience. Not only was this the nicest building I had ever been in, I had been fantasizing about the moment for months. The entrance foyer, the library, the multipurpose room, the auditorium, the gym, the cafeteria, the theater, the weight room, the pool—I had studied pictures of each in the introductory brochure, and when I saw them in person, they were already familiar. On the very first day, I felt like I was home.

The teachers, staff, and administrators were so pleasant, going out of their way to reach out to each student, to introduce themselves in the hall and ask questions, that I wasn't sure if this was a high school or a cult. Not that it particularly mattered to me; if the Prep was a cult, I was ready to shave my head, cut off my left pinkie, and start recruiting newbies.

The subject matter of the Pre-Eighth program was challenging, but I could handle English and math. Gym was a different animal. Unlike the high school, which was boys only, the Pre-Eighth grade program was co-ed. And I had little interest in exposing my athletic shortcomings in front of girls that I had only just met. While I could acquit myself well enough on the basketball court (catch the ball, dribble no more than once, pass the ball, cheer on teammates, don't cry) and was decent enough at volleyball because none of us really knew what we were doing, the weekly pool session was troublesome.

It wasn't that I disliked swimming, but doing so without a shirt on was a nonstarter. I could get away with wearing a T-shirt in the pool on family vacations because of the old "I sunburn easily" excuse. And this was true—when I was younger and carefree, I'd spend the whole day swimming in the pool shirtless, and, despite the use of sunscreen, I'd then spend these summer nights with massive sunburn and a bottle of refrigerated aloe, a can of Solarcaine, and a collection of cold, damp towels within arm's reach, staying in the motel room all by my lonesome while the rest of the family went to the Wildwood boardwalk for rides and games and fun without my lobster-red self. As I got older and wiser (and fatter), I always stuck to a white T-shirt while swimming, the ultimate protection from both sunburn and insults about my burgeoning man-boobs. But the Prep's pool was indoors, so there would be no sunburn excuse. Instead, I could only thank the Lord above when the gym instructor allowed me to forgo each pool session because I claimed an allergy to chlorine. Nasty stuff, that chlorine. Really makes me rashy and, uh, allergic.

The building and the classes and the so-nice-they-gotta-be-up-to-something faculty and staff were great. But it was my fellow classmates that really got me hooked on the Prep. These kids were different from my friends back in the neighborhood. I'm not implying that they were better. No, sir. They were not very likely to buy mice from the pet shop and release them into the wild after giving them some gin, or to pelt Ronny "Stinkin'" Lincoln's car every Wednesday for a month with rotten oranges, or to do any of the other fun shit that my buddies and I did at home. I bet if they saw their grandfather get into a fistfight with a friend's grandfather outside a bar over a televised horse race, they'd shit themselves and run home to mommy rather than shrug it off and continue lighting firecrackers. Even by my standards, they were a little soft.

But these kids were, in many ways, much more like me than my other friends. We were a collection of nerds made to feel like outcasts at our elementary schools because we liked reading books or doing our science homework. And now, for the first time, we were in an academic environment in which we were free from bodily harm and able to let our nerd flags fly. We reveled in the opportunity and experience, free to speak up in class, talk to the teachers after school, discuss with other students the merits of *The Fellowship of the Ring* vs. *The Two Towers*.

While some of the students were almost *Rain Man*-ish, as if they were so weighed down by their massive intellect that they had no room or energy to learn how to communicate effectively, some of us geeks were quite high functioning (I'd like to put myself in this category). The same was true for the girls. At one end of the spectrum were the super-nerds who could recite

Pride and Prejudice from memory and spoke French fluently but couldn't talk to a boy if he asked, "Can I borrow your pen?" And at the other end of the spectrum was Shannon.

Shannon looked more suited for Southern California than a summer school in North Philly: blonde, blue-eyed, tan, with a smile so bright it sparkled like waves breaking in the Pacific during sunset. That she smiled at strangers, didn't say "fuck," and didn't regularly pee between parked cars made her unlike any girl I'd met before. What's more, she was just so charming. She talked to boys, girls, friends, teachers, kids, and adults all in the same way. She was as comfortable standing in front of a group as she was in a small circle of people. She was polished, perfectly.

Shannon McNally (so Irish! my family would be so happy at our wedding!) pulled this off without being obnoxious and while seeming to be completely unaware of her charms. This only made her more alluring. She didn't come from a fancy private elementary school, nor was she a blue-blood that summered in Stone Harbor or Avalon or other exclusive Jersey Shore towns. No, she was the daughter of an English teacher: humble roots not unlike my own.

As her last name was McNally, she sat in front of me in homeroom that first day. Once I got over the shock and good fortune of being able to stare at this gorgeous girl without creeping her or anyone else out, I prepared to make my move. Should I make fun of the teacher's choice of tie? Or maybe "accidentally" drop my pen near her school bag? But it was unnecessary.

"Hi! I'm Shannon. What's your name?"

When she spun around in her chair to introduce herself, her hair swung over her shoulders like she was in a goddamn shampoo commercial; she was all yellow hair and white teeth and blue eyes and tan skin. I melted.

From that moment, Shannon and I were inseparable. While I attribute much of my good fortune to having a last name starting with *M*, putting me in geographic proximity of her for several hours each day, I think Shannon liked me because I could listen and I could talk, two skills that most thirteen-year-old boys lacked. Maybe because I came from a family of talkers, maybe because my mom was such a big presence in my life, or maybe because I didn't have disdain for girls the way most boys that age seem to have, I never had trouble talking to or befriending girls, not just at this age, but when I was younger and throughout high school as well. My communication with members of the fairer sex grew so frequent that for my fourteenth birthday, my mom sprung for my own phone line and telephone number so I would "stop clogging up the goddamn phone."

My crew of buddies—Phil, Vic, Floody, Jimmy the Muppet, Screech, Doc, Kruzer, Brown Eye—gave me shit for hanging out with and talking to girls, criticism that I brushed off and refused to engage. I thought to myself, Why not? What is the downside of being friends and spending time with girls? Isn't this a no-brainer? All the girls that my buddies would talk *about*, I'd talk *to*. What was so wrong with having an abundance of girl friends? The problem, I would learn, was that little space between the words. I had *girl friends*, meaning companions of the female gender, as opposed to *girlfriends*, meaning companions of the female gender who kiss you with their mouths open.

Much of my adolescence was spent trying to understand why I could not close the space between the words. With Shannon, however, I got it: this was Shannon we were talking about. Like others before her whose significant stature allowed them to go by only one name—Madonna, Cher, Liberace—Shannon was the "it" girl of the Pre-Eighth program. Typically, there were only two things I could offer a girl I had a crush on. The first was that I was smart. This was a flimsy offering at best; there were a number of connections that needed to be made in the mind of the crushee between me getting a ninety-seven on a religion test and me being successful later in life and providing her with a nice house and an ample clothing budget. Besides, every guy in this program was smart, which negated this positive. The second was that I was funny. But as I already knew, funny only got me in the front door—and never down into the basement make-out room.

I couldn't do athletic (see basketball expertise, above) or tough (my record in the few schoolyard scraps I'd had was 0-1-3—or 0-2-2, depending on whom you asked), two other qualities that most girls found attractive. And while I didn't consider myself not good-looking, in hindsight I realize that I was particularly ugly during this period in my life. Because I knew I was going to be among the poorer kids in the program, I overcompensated by dressing well. Or what I thought was well. While by any standard the early to mid 1990s was not a great period for fashion, this was especially true if you were a South Philadelphian who purchased all of your "nice" clothes from a "stylish" shop known as Gigolo's, whose target customer base was the IZOD and IROC-Z crowd. So while the rest of the kids wore sports T-shirts or other normal twelve- and thirteen-year-old kid clothes, I came to class each day decked out head to toe in Z. Cavaricci, eager to prove that my family had plenty of money—just look at my awesome clothes! Except they were not awesome. Not then, not now, not ever. And I ended up not looking like a student, but like a student's Italian-American lesbian aunt who dressed really, really tackily.

The author (right), dressed as a middle-aged Italian-American lesbian aunt, with friends from the Pre-Eighth program.

And none of this takes into account that Shannon showed no inclination to look for a boyfriend. This didn't stop many of the guys from acting like the program was a seven-week talent show in which the winner was awarded Shannon's heart. At every opportunity, the males paraded in front of her or otherwise tried to get Shannon's attention through words or acts that could be boiled down to, "Hey, look at what I can do well, Shannon! It is my sincere hope that this increases your opinion of me!"

Add it all together, and though I had a crush on her so heavy that at times I could barely stand, I knew it was a lost cause, an endeavor not even worth exploring. I was content with my consolation prize, her attention and her friendship. Shannon would tell me about her hopes and her dreams and her plans for high school and beyond, and I'd smile and nod, happy to be in her presence, wondering what her hair smelled like (my guess: cinnamon in the fall and "crisp breeze" for the rest of the year).

Though I knew that Shannon would not attend the Prep, the experience of meeting her was a big reason why I decided that I wanted to go to high school there. Not only was the Prep everything I imagined it would be, but I had met and befriended a stunning, charming girl the first time I went there. Win-win.

The Pre-Eighth program was the easy, fun part. If I wanted to go to the Prep, making it happen would be a little more difficult.

In Classroom 107 at the Prep, on the November morning of the high school entrance exam, the proctor, taking roll call, called out, "Conrad Benedetto?"

Conrad Benedetto? What the hell kind of name was that? In my part of town, if you weren't named after a saint and your last name didn't have a whiff of Gaelic or end in –ski or –sky, you were considered a weirdo. (There was no Saint Jason, but there were three other Jasons in my class of eighty students in elementary school, so *Jason* was also acceptable.) But *Conrad Benedetto?* Was Otto van Mascarpone also in the room? How about Wolfgang Amarillo?

That I wanted to attend the Prep was met with mild disappointment from my dad and grandfather, who wanted me to be a Neumann man. I could understand my grandpop's wish that I continue the tradition he had started almost fifty years before, but I found it more interesting that my dad wanted me to go to Neumann. By his own admission he didn't remember most of his senior year, and he was not exactly a booster for the school, regaling me with tales of high school hijinks that usually involved the heavy use of drugs on school grounds. I think his interest in my going to Neumann was related more to the Prep's oversize price tag than to his desire to see his son at his alma mater. As for grandpop, he had about a dozen male grandkids. One of them would go to Neumann for sure, so he'd get over it.

My friends were a little more vocal. That I didn't want to go to the same high school as almost all of them led to various rebukes and insults along the lines of "fancy boy" to "what, you're too good for Neumann?" It was just a school, I countered. It was no big deal. Besides, it wasn't like I was moving or anything. We'd still hang out all the time.

In contrast, my mom was thrilled and supportive. Though the school's annual tuition was equal to about two-thirds of her salary, it didn't matter to her. If a requirement for my matriculation at the Prep was that she had to wrestle a lion, she'd show up an hour early to stretch before the match, manhandle the lion, and then ask for another.

The summer over and now back at OLMC, eighth grade so far had been enjoyable because it had been unlike any of my previous years of school. For the first time, school did not require rote memorization and recitation ("Name the beatitudes! Who was the nineteenth president!"). Rather, there was more interaction between teachers and students. Mr. Bruno, the prototypical cool male teacher who had previously talked to us about being nicer to young ladies, did not make us recite presidents eleven through twenty, but asked us who we thought had been the best president and why—and what made a "good" president, anyway. In religion class, while the Catholic party line was that the devil was a red dude with a tail who lived underground and was a real asshole, Ms. Peterson taught us about other religions, which many of us didn't even know existed ("You mean, there are people who are not Catholic?").

English class was the most different. Long accustomed to diagramming sentences and reading boring books, Ms. Flynn blew our minds when she said we could read any book we liked and keep journals about them that we would then pass around to our classmates. So I could write in my journal a note to Vic about anything I wanted, as long as it mentioned something about the book I was reading. While it may sound like a recipe for disaster, it actually worked. Sure, my friends and I might write long journal entries to each other embedded with private jokes that bore no relation to what we were reading, but we also read books—a lot of books—and discussed them in our journals. Ms. Flynn approached the writing portion of language arts class in a similar way. We could write whatever we wanted—poems, essays, short stories, plays. Then we'd break into groups every few classes to workshop what we'd written. One notable work was Mark Hutton's tour de force short story, "The Day I Spilled Hot Chocolate on My Privates," in which he wrote about the day he spilled hot chocolate on his privates. If this isn't in development at Fox right now, the entertainment industry really is dying.

But despite being turned on to reading, writing, and learning like I had never been before, no amount of love of learning would get me into the Prep. It all came down to the entrance exam and one number: ninety-eight. Ninety-eight was the minimum I'd need to score on the entrance exam to qualify for a scholarship. If I got the scholarship and also a work grant and some financial aid, my family could make it work.

I was nervous for the spelling bee, knowing that Brutus was on the line. But going into that bee, I was confident and didn't feel so much pressure, because I knew a top-seven finish would get the job done. With the Prep entrance exam, I was much less confident. Ninety-eight did not leave a whole lot of room for error. I didn't know how many questions there would be, only that the test was multiple choice and would cover English and math. But if it were one hundred questions, and I got three wrong, no Prep. My high school career—and my entire life—could hinge on one single question. When I lost the bee, my consolation was that my quest for a dog was not over. While I was not going to get my Brutus anytime in the near future, I could get a dog in a few years, or maybe when I went to college, or certainly at some point in my life: when I was an adult and had my own house, I could get dozens of dogs if I wanted to. With the Prep entrance exam, it was different. I either got the scholarship and went there or didn't get the scholarship and went to another high school. There was no "maybe later . . ."

The lesson I took from the Brutus experience was that life is not a Disney movie. When it comes to competition, you fuck up, you're done; someone else is there and ready to win what you've lost. While my parents promised that they would do their best to find the money even if I didn't get the scholarship, I wasn't going to bankrupt the whole effing family just so I could go to a private high school with a pool and Latin classes. I needed that ninety-eight.

You know how trauma victims black out much of the circumstances of the accident that injured them? "I remember being on the train tracks, I remember seeing the train's headlights, and I remember waking up in the hospital bed."

I remember Classroom 107. I remember Conrad Benedetto. And I remember getting a letter in the mail some time later congratulating me on my Ignatian scholarship to Saint Joseph's Preparatory School. And with a letter came a brochure with smiling, fresh-faced student council officers on the cover, whose contents I could recite from memory.

football line

For the first dozen or so years of our lives, our decisions are made for us. What we eat, what we wear, where we go to school and the type of schooling we receive, what God we do or do not worship, what kind of movies and TV we watch, what music we listen to, where we travel—we're just along for the ride and accept what may come. We produce little, consume everything, and that which we consume is provided by our parents or the authority figures in our lives.

But then, slowly, things start to change. Right around puberty, we begin to assert greater control of our lives and start to make those decisions previously made for us. Now you can wear whatever you want, even if your parents forbid you to, by, for example, changing clothes at a friend's house before school. No sweets? You can stop at Dairy Queen every day after class and no one will be the wiser. If your lame-ass dad bans all gangsta rap in the home, you can always listen on your headphones. Raised Evangelical Christian but thinking the whole pagan thing is more your speed? There are plenty of message boards and meetings (presumably secret, presumably in the woods) to explore. Indeed, from my understanding (which consists of my thinking about this for the first time as I write and making shit up as I go along), puberty gives a child the opportunity to assert his or her individuality, often in defiance of his parents' wishes.

It is around this time that the first major decision on which we have any input presents itself: where to go to high school. Most people would agree that the consequences of this decision are great and that it involves a traditional milestone in one's life. Of course, our parents have a significant amount of input on this one, too. Cost is a major consideration, as it was in my case. But at the same time, it is difficult for any parent to force his or her child to attend a high school he doesn't want to attend. So this is a kid's big chance to speak up and be heard.

After high school, the other major decisions are all ours. College or work? Or maybe a year traveling through Europe, having sex with strangers, and "discovering" oneself? After that, is it time for a real job? Or maybe grad school? Or maybe temporarily living with the 'rents while you tackle the next great American novel? Then, maybe, the process of spouse-choosing and marriage? Or a commitment to staying single? To procreate or not to procreate?

These are all heavy, heady choices. But if Malcolm Gladwell has taught us anything, it's that while the big ones are important, the cumulative effect of the little decisions are just as significant, if not more so, than those big-picture decisions. Or that decisions that at the time seem inconsequential can later turn out to be quite impactful and truly define our lives.

(He wrote a book about something like this, right? It certainly feels like he did. If he hasn't, he's probably working on one right now.)

On our first day at the Prep, we were placed into one of seven different homeroom sections, in which we would have the majority of our classes freshman year. The sections ranged from 1A to 1G and were based on what each student received on the entrance exam, which was supposed to correlate to his academic ability or promise. This was not explained to us, but was made evident by lunchtime. The kids who sat together in cafeteria from 1G looked like the smartest kids in the class: the über-nerds. While they weren't eating with microscopes and protractors splayed across their lunch tables, it wouldn't have surprised me if they were *talking* about microscopes and protractors. The kids from 1A were at another section of tables across the room. While you wouldn't call them *dumb*, they were definitely boisterous—yelling and carrying on, placing french fries on their crotches as if the fries were their dicks. One guy blew the room open when instead of a french fry, he put a much larger chicken finger on his crotch. Solid move.*

My section was 1F. This made sense; I'd gotten the ninety-eight on the entrance exam, and was placed with a few other 98ers and scholarship kids. I was able to decipher this from our morning sessions, but that was about all I could pick up. The near-mystical quality of the Prep that I had felt throughout the

---

* I wonder if the Prep still uses this nomenclature to differentiate the sections of classes, or if a more politically correct nomenclature is used, like the "Green Group" and the "Red Group." Similar to "MG," what were administrators thinking back then—only ten or twenty years ago? Fifty years ago, did they have "Group Smart Kids" and "Group No Chance"?

Pre-Eighth program and prior to attending was absent on that September day, replaced instead with first-day jitters and the all-business attitude of the faculty and staff now that school was in session. In the morning, we were herded from class to class for mini-periods—a full day's walk-through of our roster in half the normal time. We were given syllabi (and were informed that *syllabi* was the correct plural form of *syllabus*), and each teacher explained his or her class and what would be expected from us. Then lunch with the geeks and the french fry dicks and everyone in between. After lunch, the afternoon session.

The afternoon session brought together all two hundred–plus freshmen for a series of presentations from the administration. After we were seated in the auditorium, the first person to speak to us was Father Spencer, the president of the school. A tall, kindly man, he had traded in the brown robes of his doppelganger Obi-Wan Kenobi for a clerical collar and the Society of Jesus. Instead of preaching to us about the Force, Father Spencer spoke about the importance of being a "man for others," a Jesuit ideal. I think this had something to do with Saint Ignatius and social justice and whatnot, but I was feeling pretty sleepy after a heavy lunch of a surprisingly delicious cheesesteak from the cafeteria, so I was zoning in and out.

I perked up, however, when Father Spencer spoke about how the road ahead was going to be difficult, academically speaking. He said that we should expect to have no less than three hours of homework every night. This would seriously cut into the time I had set aside each evening for video games and playing with myself. Then Father Spencer paused for a moment and told us, "I want you to take a moment and look at the student on your left. Now, look at the student to your right. The odds are that one of you will not be here by the end of the year."

It was a scare tactic. And it worked. You could hear a dozen separate whimpers from the audience, and if you listened closely enough, you could make out the faint sound of someone peeing himself somewhere in the back right of the auditorium. That I couldn't hack it at the Prep was something I'd never considered. To not just fail a test, or even a class, but out of the high school? No way. This type of defeat was not an option; I could not slink back to Neumann, a failure and flame-out, to the unending ball-busting of my friends. No way. If it took six hours of homework a night, I would make it work.

After that zinger, Father Spencer gave way to Mr. Kearney, the dean of students and disciplinarian. If Miss Piggy were a man—a really, really pissed-off man—and one bad motherfucker, she'd be Mr. Kearney. Country strong, with heavily pomaded hair and thick glasses that gave his eyeballs a bulging appearance, Mr. Kearney told us how things were going to go. With his face alternating between the various shades of red and pink you'd find in a Hallmark store in early February, he said we were to adhere to the dress code (blazer, shirt, tie, slacks), we were to be on time, we were to be respectful, and we were to be Christ-like. As he said each of these things, he pounded his cantaloupe-size fist on the lectern, showing off a class ring with a purple stone larger than either of my balls. With his multiple Christ references, he struck me as some sort of goddamn Catholic vengeance warrior; I could see him in a fit of rage suddenly breaking a chair over a student and screaming "May Christ have mercy on your soul!" in Latin.

The penalty for misconduct was called "jug." This was the Prep's version of detention. Mr. Kearney did not get into the origin of the word, but later we learned it was one of two things: the acronym for "justice under God" or a derivation of the Latin word *jugum*, which meant "yoke" or "burden." So if you acted out, instead of getting held after class, you were either going to be served justice under God or possibly strangled, throttled, and yoked as would a beast of burden. Either way, so much scarier than *detention*.

The one-two punch of Father Spencer saying that many of us wouldn't survive the year and Mr. Kearney's "God may own your soul, but your ass is mine" speech left us reeling. When Mr. Kearney introduced Mr. Jacobsen to speak to us next, there was tentative, stunted applause, as most of the audience was too shocked or frightened to know whether to clap. I was almost literally shitting myself; in addition to the stress and fear I now felt, the lunchtime cheesesteak had begun to set off a series of disturbances and warning shots in my belly.

Mr. Jacobsen, the director of admissions, who was already familiar to all of the students in the auditorium, tried to lighten the mood by opening with, "And now, the fun stuff!" His short talk focused on student life and all the exciting extracurriculars the Prep had to offer. He explained that representatives from the various clubs, organizations, and sports at the Prep were going to speak to us about their respective activities. After the talks were over, we'd be free to go home or stick around to learn more about the activities.

I wasn't planning on joining any clubs; not right away. I was interested in a few—some of the mega-clubs like Students Against Drunk Driving (SADD) and the Community Service Corps (CSC), which boasted dozens and dozens of members, as well as maybe the newspaper or the yearbook committee—but I had to maintain a certain grade point average in order to keep my scholarship and I didn't need any club taking up too much of my time and causing my grades to tank, which was at the front of my mind after Father Spencer's speech. And besides, I could always join whatever club I wished later in the year or sophomore year, once I was sure I could handle the academic load.

There was only one fall activity that I was really interested in: football.

There was nothing in my past or in my personality that indicated I would want to play high school football. While I wouldn't say that I was hopelessly unathletic, many others, including members of my family, my closest friends, and the people who knew me best, would. I generally preferred sitting and talking to moving and not talking. Sports—and really all physical activity—were a necessary evil, an unavoidable means to an end. (I want ice cream; I must walk to get ice cream.)

The author (second from right), taking part in one of his few athletic pursuits—in a matching sweat suit (I know the picture is in black and white, but you should know the sweat suit is a gorgeous turquoise color. I was very big into bright blues at this time).

And yet, I was drawn to football because of the simple fact that I was big.

Big, chubby, fat, whatever you want to call it: that was me. As a freshman, I was already five-eight or five-nine and over 170 pounds. This did not make me freakishly big—there were bigger kids my age, for sure—nor was I one of those kids who showed up the first day of ninth grade with a full beard and a credit history. But I was still a large kid, always toward the back when we lined up shortest to tallest, often the subject of insults about my weight. I was big, yes. But I was soft. And that's where football came in.

What others saw as a heap of flesh with a deep, abiding fear of lightning and barely enough coordination to tie his shoelaces and breathe at the same time, I saw as a fount of untapped potential. It was not that I was unathletic. It was that I needed athletic direction, coaching, support. My dad, as fathers often do, tried to give me some of this coaching. But my father was not an athlete as much as he was tough. Though he, like his four brothers, played football in high school, my dad's toughness was not gained on the gridiron but through hard work as a longshoreman and the occasional fisticuffs after a rowdy night out drinking. He and I threw the old pigskin around every once in a while, but we were not going to head down to Coley's to drink whiskey and pick a fight.

Though my dad and I were on opposite ends of the toughness spectrum, I did have his genes, which was a plus. Perhaps if I got involved with football, with all its mandatory practices and weight room sessions, I could make myself into something—I could realize this potential. What if I kept growing and getting taller and turned some of this baby fat into muscle? And what if, as I got older and fitter, I'd grow into some of my dad's toughness? I'd be fucking unstoppable, that's what if.

I couldn't play youth league football because I had been too big. In high school freshman football, I could use my size to my advantage, because the division of teams was based on class year rather than weight. I could also meet people while playing football, because I wasn't going to join any clubs. And unlike clubs, if I joined football during sophomore year, I'd be too far behind. I had to start with practice number one, with the rest of the freshman team. And it was not lost on me that playing football might help with my titty-feeling ambitions, which were arguably as important as getting in shape and making friends. If you were a goddamned American, whether in Oelwein, Kansas; Lancaster, California; or South Philly, you knew that guys who played football got the girls.

The first person to speak was Coach Marks, the Prep football coach. What was remarkable about Coach Marks was how much he *looked* like a "Coach Marks": Central Casting could not have done better. I wondered if he ate dinner with his family with his whistle around his neck, if he showered with his clipboard, whether he yelled "get low! get low!" at his wife during sex. Dressed in a polo shirt, he projected a toughness on par with that of Mr. Kearney, but he was less formal and more charming. He spoke about the team, which was good but not a football powerhouse, with a passion that was intoxicating—all sorts of manly shit. The teachers and staff would make us spiritually and academically strong, he said, but football would make us physically strong. I had already made up my mind and didn't need convincing about playing football. But if Mr. Marks had been the fencing coach, I would have jotted down "Get mask thingee and bendy sword" in my notebook about a minute into his speech.

While Coach Marks spoke, I began to panic about my gastrointestinal distress. I had always been something of a nervous pooper. Perhaps it was all becoming too much for me: it was the first day of school, we had had all those mini-periods and went quickly from class to class, I had Father-Wan Kenobi telling me I might flunk out and Mr. Piggy scaring the crap out of me, I was hot and uncomfortable in my jacket and tie, and we still had a number of speakers to get through. As I was seated in the middle of the row, I couldn't get out of my seat without causing a major disturbance. And anyway, I'd look like an asshole, walking out when someone was speaking. So I had to sit there and deal with it, suffering, sweating, and wondering if the kid sitting next to me could hear emanating from my stomach what sounded like a series of large meatballs being dropped into a pot of thick gravy.

Finally, mercifully, after a number of speakers and a seat-shift/sphincter-clench every few seconds, Mr. Jacobsen came back to the podium, thanked us for our time, and told us we were free to leave or to head over to the multipurpose room, where we could meet club and sports representatives and sign up. That sounded cool and all, but it was time for the reckoning. While others were slow to get out of their seats, sighing, stretching, and relaxing after the long session, I jumped up and over a number of students in my row, then ran straight up the aisle and right out the door.

Suffice it to say, the poo was everything that I dreamed it

would be and more.* It was made even better because I had the entire bathroom to myself. There was a bathroom just out-side the auditorium, but I figured that would be flooded by my fellow freshmen, who might get alarmed if there was any grunting, punching, or moaning involved (likelihood of any of these happening: over 85 percent). So instead I went to the main bathroom upstairs by the locker room, thinking that if I made it through an afternoon of sitting in the same seat and listening to speeches, I could make it a few additional feet and seconds to enjoy a more spacious—and empty—bathroom.

I relaxed and took my time, decompressing from the long day. Looking back, it may have been one of the happiest moments of my life. I had made it through my first day of high school! At the Prep! And without shitting myself! Well, I'd *almost* made it through my first day. There was the small matter of signing up for football, but that would only take but a moment, and then I'd be on my way home, day one in the books. I was now here, where I belonged. Not so much in the bathroom stall, but, you know, the Prep.

After a good while, I left the bathroom and walked to the multipurpose room to join many of my classmates and to find the line for football sign-ups. This was easy to do, because the football line was the longest by far. There were two freshmen by the literary journal line, a handful around the debate club, and about half the freshman class in the goddamn football line, waiting their turn to speak to the coach and the team captain and to sign up.

---

* My God, I'm proud of that sentence.

And in this moment, my decision had to be made. What did I want more? Did I want to play freshman football, to get myself fit, and to transform myself into a lean, mean killing machine? Or did I want to *not* wait in that super-long-ass line and go home instead? I thought about it for a few seconds. And maybe a few more seconds. Then I realized it was the latter. By a landslide. I walked out of the Prep and to the bus stop. Day one in the books.

Could I have signed up for football later? I suppose. But that would have required going out of my way, finding Coach Marks, introducing myself, and explaining why I didn't sign up after the first day of school. I couldn't tell the truth, as I wasn't sure how well "um, I pooped for a while and then didn't feel like waiting in line" would go over. If I lied and left out the poo and said I was in another line, I was sure he would retort, "What, you couldn't wait for a few minutes in the football line? What kind of commitment is that?" If I couldn't *stand in a line* for ten or fifteen minutes, how could Coach Marks and my teammates be expected to count on me on the field of battle? He'd have every right to chew me out and kick my noncommittal ass out of his office right then and there, embarrassing me in front of the whole team in the process.

Once I missed the sign-up sheet window that afternoon, it was over for me. I did not play football freshman year—or in any year in high school. I didn't play any sports in high school at all. As a matter of fact, I failed gym my sophomore year because I didn't meet the minimum requirements of twenty hours of organized physical activity. For those who didn't play on a sports team, there were intramurals—basketball, volleyball, and ping-pong among them—through which even the least athletic kids could get their twenty hours of activity in. But sophomore year, for reasons that escape me, I didn't hit the threshold. I had to come to summer school for one day to shoot hoops by myself for four hours, the only person in the entire school required to do so. I'm pretty sure that before me, no one had ever failed gym. And when I did, they realized they didn't have a precedent or a plan for what to do with me. ("Just tell him to come in for a few hours on Monday and, uh, shoot baskets or something.") Then I got my F changed to a P on my transcript and took the bus home. It was not my finest day.

In hindsight, I look at that poo as one of the defining moments of my life. A simple bowel movement not only drastically altered my high school career, but everything that came after it. What if I had not gone to the bathroom and instead signed up for and played freshman football? What if I had been good and enjoyed it? I won't go so far as saying that I'd be in the NFL right now, getting blow-jobs in my limo while drinking brandy from a chalice. But if I'd spent four years working out year-round, maybe that would have instilled in me a work ethic or a passion for physical fitness. Instead, during high school, I accumulated pounds the way one might accumulate rental properties in a down housing market: hastily, en masse, with an almost unconscionable greed. If I had baby fat as a freshman, I had middle-age, diabetic fat by the time I graduated.

In fact, even as I write this, I have no discernable muscle tone on my body. Sometimes I go for a run, as long as we understand "sometimes" to mean "so rarely that I could say 'never' and get away with it," and "go for a run" to mean "put on all my gym gear, go to the gym, walk around the floor of the gym sipping water, and walk home." (I will say that I am very good at making workout playlists, for whatever that's worth.)

If I had played football, I probably would have made and been friends with the guys on the team, many of whom went on to become the BMOCs at the high school, going to and throwing the coolest parties, making out with the cutest girls, drinking the beeriest beers. Instead, while the friends I made in high school were good, funny guys, we could not exactly be thought of as pussy killers. We talked to girls on occasion, yes, but most of them were employed by a business we were patronizing. And we went to parties and drank some alcohol, but in the case of my friends and me there was usually a very public self-urination or vomiting or similarly shameful moment involved rather than a fingering in the nearest closet.

It wasn't all doom and gloom, though. Yes, because my poo and laziness prevented me from signing up for football on that first day of class, I probably missed out on a few minor things, like high school sex and the unbridled euphoria of being brought to orgasm by another person, invites to the coolest parties, general good health, esteem from my peers, approval from my father, and pride in myself. But I did gain in other areas.

Right now, you, [your name here], are reading a book about my experiences as a fat kid in high school. I'd like to think that you wouldn't read the memoir of a teenager who was a reasonably decent lineman, had sex a few times, and graduated a member of the jocks. (Although if you're reading this, there's no telling what you'd read.) And the friends I did make in high school I count as my best friends today. So . . . that's a bonus. These things beat all that other stuff about sex and being cool and respected and fit any day of the week.

Yep. Any day of the week.

master of crabs

David "Floody" Flood and I met in homeroom in the first grade. Our big friendship moment came a few months later when I was invited to his sixth birthday party, a week before Christmas. My gift, purchased by mom but in consultation with me, was a large, heavy action figure of the WWF wrestler Big John Studd, made of the same dense rubber material out of which sturdy dildos are made.* I was pumped; I would have loved to add Big John Studd to my WWF collection.

We sat in a circle at the party. David opened my gift and said, "Oh, I have him already."

I was bummed and began to sulk. I thought I was giving the most awesomest gift of all time, and he already owned it. How

---

* I don't know if women (or men) name their dildos, but if you do, I beg you to consider calling yours "Big John Studd." Really, I can't think of a better name for a dildo:
GIRL 1: "What'd you do last night?"
GIRL 2: "Oh, I had a great Friday night. I stayed in and drank a bottle of malbec, and then Big John Studd came out and shit got real."

was I supposed to know that? Six-year-olds don't have registries. I drowned my sorrows in a big bowl of popcorn.

After the gift-giving portion of the party was over, David approached me and thanked me again for the gift (likely prompted by his mom, Eleanor, who couldn't have helped but notice my sulking). He said that anyway, since he already had one Big John Studd, maybe he could keep the one I got him and we could melt the old one over the stove, just to see what happened. I smiled. He and I have been best of friends since. (Although Eleanor prevented us from doing any Big John Studd melting.)

A high-energy guy, David possessed a tremendous amount of street smarts and always seemed to have a plan or an idea or a hustle going. He and I had a number of adventures and ventures growing up, including a short-lived business selling fireworks that, like our Big John Studd–melting idea, was ultimately quashed by his mom. Such a setback did not stop David from being the hardest-workin' kid on Second Street.

In the fall of our freshman year of high school, fate smiled upon David, and he got the coolest job that any sports-obsessed thirteen-year-old boy could get: he became a ball boy for the Philadelphia Eagles. He had a friend who worked as a ball boy but had to quit, and the friend offered the job to David, which he accepted without a second thought. Now he would train with the team, work all their home games, and spend all summer with them in the camp. With the Eagles! The NFL team! The one we watched on TV and talked and complained about! David would be in the locker room with them, and on the field during games! It was a dream come true.

And David *already* had the coolest job of anyone I knew. While the rest of us were delivering papers or shoveling snow or scooping ice cream, David worked in a bar. A real, actual drinking bar. It was called Mick-Daniel's and was owned by his uncle Mike, who had bought the bar a few years before. Even in a neighborhood where there was a corner pub every other block, Mick-Daniel's stood out as *the* destination. When the wedding receptions ended, when the graduation parties were over, when the block parties had run out of booze, it was over to Mick-Daniel's to keep the fun going. It had been the neighborhood hot spot for years, even under the previous ownership when it was called Friends' Tavern (pronounced FREN-zees). The bar had a place in my family lore, as my parents had gotten engaged there. My dad had the cook, his friend Dan, stick the engagement ring in a bowl of cheese fries ("but not the cheesy part"). My dad was also arrested at Friends'. Twice. Though not on the night he proposed to my mom.

Despite my father's apparent willingness to break laws within or in the vicinity of the bar, during lunchtime and the early evening hours Mick-Daniel's was a family-friendly place, on account of a kitchen that served delicious and upscale (by South Philly standards) pub food. This is where David worked—in the kitchen, as a dishwasher.

Dishwasher is on the bottom rung of the kitchen food chain, but that was not important. David had a job in a bar. And not just any bar, but *the* bar. I'd eat at Mick-Daniel's with my family and when I looked up to see David, peering out from the kitchen, talking to the chef or filling up ice behind the bar while joking with the bartenders, I was jealous. David would meet up with us on a Saturday morning and he'd stoke our jealousy by saying, "Man, you wouldn't believe how crazy the bar got last night." He actually got paid to hang out in a bar. It was his best hustle ever.

But when the Philadelphia Eagles come calling, you drop what-ever you're doing and run to them. Fast. Hell, if they'd asked him to drop out of school, I'm sure David would have considered it. David had to leave Mick-Daniel's, and he needed a replacement to take over his dishwasher/bar-hanger-outer job. So he asked me.

On the one hand, it made a lot of sense. I was still delivering papers, but now that I was in high school, it wasn't an ideal gig. The papers had to be picked up right after school and delivered by 3:30 p.m. When I went to elementary school in the neighbor-hood, this had not been a problem. But now I went to high school in North Philly, which meant I had to take a bus home every day, which meant that the earliest I could get home was 3:30 p.m.

Because I had been working the paper route for years, my boss, Franny, allowed me to deliver a little later than the other paperboys. But the job still left me no flexibility. I wasn't going out for football, nor was I planning on joining any clubs. But what if I wanted to stick around and do homework in the li-brary? What if I wanted to hang out after school and shoot the shit with my new friends? What if (gasp!) I got jug? I could do none of these things with the paper route.

If I took Floody's job at Mick-Daniel's, I could drop the paper route and would only have to work the busy Thursday and Friday dinner hours and cleanup (from 4 to 10 p.m.). Just two days a week instead of five. Also, I'd make more money and have more flex-ibility. And me working at Mick-Daniel's! (!!!) Um, yeah, I might have been interested in that. More money, more free time, and an immeasurable boost to my cool factor. Yes. Yes, please.

I must pause for a moment to clarify the reason my friends and I thought working in a bar was cool. It was not because it brought us closer, if only geographically, to drinking alcohol. Some of us had started to drink at that point, but drinking took place under the bridge, in the schoolyard, or in other barren places with dark shadows. There were no thoughts of stealing drinks or being able to drink with the staff: both were out of the question. We were thirteen years old. This was Second Street, not Russia.

And though many of my friends had been in the bar before, it was to eat with their families. To work there meant you'd be there alone, unsupervised by a parent. And it meant that you were also contributing to the bar. You were a vital part of an organization that employed a select group of people and relied on those people to keep the business running successfully. That this organization just so happened to serve alcohol was a bonus.

While some bars could be dreary, depressing places where hopeless men and women measured time by each well vodka-rocks until death came knocking, Mick-Daniel's was a magical place, a wonderland where people came for one of two reasons: to get happy or to keep the happy going. To work there was to be a part of this happiness, to actively contribute, even in a small way, to this happiness. Being an employee at Mick-Daniel's would offer the same job satisfaction as, say, an elf working in Santa's workshop has, knowing that the toy he was working on would bring great joy to a child somewhere in the world.

I took the job. But before I could start, I had to quit the paper route. I told Franny that I had to leave to pursue another opportunity and thanked him for six years of a good job, telling him that I appreciated the opportunity and that I hoped our paths might cross again in the future. After I finished my little speech, he said, "You got a brother?"

"Yeah, I have a younger brother."

"Can he do it?"

I thought about my brother Dennis, then in fifth grade. It wasn't a demanding job. You needed a basic understanding of math and geography, as well as the ability to walk a few blocks with the somewhat heavy bag of papers over your shoulder. But Dennis was at the peak of his morbid obesity. When hanging out at my house, my friends and I would make Dennis sit in the middle of the couch in the living room and ask him to get up from his seat without using his arms. Try as he might, he could never do it: his fat ass would be stuck in the middle of the couch, unless he pushed off with his arms. If he couldn't sit up on his own, I had reservations about his ability to carry a heavy bag in mid-July.

"Oh, no. No, he can't do it."

Dennis. Built for meatball eating, not paperboy-ing.

My five-year connection to the *Philadelphia Daily News* now over, I reported for work at Mick-Daniel's that Thursday. But before I was allowed to work on my own at the new gig, I had to shadow Floody for two shifts. He told me to meet him at the bar at 4 p.m. I requested that we meet at his house and walk to the bar together. While I'd been there before, I'd never walked in without an adult.

Together, Floody and I walked through the hallowed portal and then right into the kitchen. There I met the waitresses, his mom Eleanor, and Sherry, a friend of my mom's whom I'd known for years, as well as the prep cook, Karen, and the chef, Matt, all of whom made me feel at home by immediately making me do shit.

David showed me how to wash the dishes and his little tricks for doing them more quickly, took me to the walk-in freezer to show me where the mozzarella sticks, chicken fingers, and beef for the cheesesteaks were kept, and demonstrated how to empty the grease from the grill. An hour after taking me through the basics, David left. An hour and two minutes after David took me through the basics, I was rooting around the walk-in freezer because the kitchen was out of onion rings.

Over the course of the next year, I became a master at the art of washing dishes. I would do the menial stuff like filling up ketchup bottles and washing the floors after the kitchen closed, but eventually, I was given greater responsibility, like prepping dishes for the chef or helping the short-order cook and even grabbing cases of beer for the bartenders when they asked me to.

The dinner hours would fly by. I'd get in at 4:00 and the next time I looked at the clock, it would be 8:30. But it was after the dinner hour when the bar really came alive. The lights dim, the bar full, the voices and the laughter loud, I'd sweep the kitchen floor, peering out like I used to watch David do, mesmerized. So that's what you did after you grew up.

Once the spring turned into summer, the owner of the bar, Mike, asked me if I wanted more work. With a summer full of nothin' planned, I told him sure, I could use more hours.

It was more hours, yes, he said. But it was a special assignment, involving long days, and would not be easy. As a matter of fact, it could be pretty scary, even dangerous. Was I still interested?

The truth was that no, I was no longer interested. I liked working and liked money. But I did not like the sound of "not easy" or "scary" or "dangerous."

But it would be difficult to turn him down. Mike was a pillar of the community: owner of the neighborhood bar, friend to all, good guy. Also, an intimidating dude, one around whom it would not be appropriate to admit fear. He was also the same generation as my dad and knew him well.

My dad being such an animal in his younger days did not make it easy on me growing up (and I use the word *animal* as lovingly as possible). Though I never saw him act like a madman, I knew all the stories. There was the time "downnashore" when he was seventeen, and he jumped head first off a pier into the bay and broke his neck, but he was so drunk (etc.) he didn't learn his neck was broken until the next day, after which he spent the rest of the summer in the hospital. There was the time when my mom first laid eyes on him: he had been stabbed and was bleeding, but he was too drunk (etc.) to notice (and no, my mom was not a nurse or

medical professional). There was the time when . . . well, you get it.*
While he often told the story of his broken neck, happy to show the
long scar to any listeners, the other tales I heard from his friends.
These friends would volunteer stories about my dad shortly after
meeting me and asking, "Who's your father?" (Because of the large
family, there were a lot of kids with the last name *Mulgrew*, and
you had to place them with the appropriate Mulgrew parent to size
them up.)

My dad the multitasker: spending time with his toddler, going for a walk, and getting drunk as shit all at the same time.

---

* Previous book plug #1: For these and other stories, please check out *Everything Is Wrong With Me: A Memoir of an American Childhood Gone, Well,
Wrong*. Available online and at fine bookstores everywhere.

The stories that I'd hear about my father from his old friends were so similar to each other that they could have been Mad Libbed:

"Your dad is Dennis? Oh, man. I remember this time I was out with your dad, and he _pooped in a shoe_ and then we found _a bag of industrial-grade Demerol_ in _an abandoned baby stroller_ and went to _a Bad Company concert_, where he _fought a brown bear_. He _lost badly_ and _took more pills_, and then we _stole an ambulance_. Man, we had fun back then."

That was my dad. And then, there was me.

A representative but noncomprehensive list of things I was afraid of would include: bugs, the garbage disposal, the dark, balloons, people with dreadlocks (yes, even the dorky white guys), that loud noise when you close a heavy hardcover book really fast, any member of the horse family, heights, sharks, the ocean, the sun, staplers, and the usual (ghosts, werewolves, vampires, mummies—all monsters, really). It is perhaps no coincidence that although I was the firstborn, it is my younger brother, the second child, that is my father's namesake.

INT. HOSPITAL ROOM—JULY 1979
A DOCTOR, having just delivered a BABY, hands it over to the PARENTS.

DOCTOR
Dennis and Kathy, congratulations. It's a boy!

KATHY
(taking baby)

Oh, Den! Look at him—he's beautiful! Should we name
him "Dennis"?

DENNIS
(a beat, takes drag from cigarette, peers at baby)
Eh, I got a bad feelin' about this one. Let's wait for the next
one.

So when Mike presented the not easy/scary/dangerous job
opportunity, I couldn't say no. I had to prove I was Dennis
Mulgrew's son, goddamn it, even if I wasn't named after him.
I could manage not easy/scary/dangerous with my eyes closed.

When I said that yes, I was still interested, Mike yelled back to
the chef and said, "Hey, Matt. I think we got our new crab man."

Many of the bars in the neighborhood had crab nights in the
summer. For ten dollars, you could get all-you-could-eat crabs
for a few hours. Mick-Daniel's was no exception. Thursday nights
were their crab night, and it was a big social night out. For one, it
was cheap and you could eat until your heart stopped, two things
Philadelphians just love. Also, many people took Fridays off in
the summer, so the crab night was the start of their weekend. Go
out with some friends, drink eight pitchers of beer, eat a hundred
crabs—what better way to kick off three days of summer fun?

I myself didn't eat crabs. I thought they were a lot of work
for little reward. Hadn't we evolved enough as a species that
we didn't have to crush the exoskeletons of our food and eat
their meat with our bare hands? There's all the smashing and
the pieces of shell flying everywhere, and you use newspaper as
tablecloth (are we homeless?) and you get shit caked under your

fingernails. It was barbaric. And I won't even get into the whole mustard-poop thing. No thanks.

But I didn't have to like crabs to be the crab man. Actually, it would behoove me to dislike the crabs, since the crab man was really the crab executioner.

The crabs were brought in at 11 a.m., when the bar was not yet open and was eerily empty. I came into work that first summer Thursday to find the five rickety crates teeming with life stacked on top of each other in my area near the sinks. They stopped me in my tracks. The stack of bushels was moving, for chrissake, full of hundreds of crabs aware that the clock was ticking, the end near. The chef, Matt, cigarette dangling from his mouth, came over, patted me on the back, and said, "Well, they're all yours!"

Matt explained that he would make the broth and cook the crabs, but I had to prep them. This meant putting the crabs into an induced coma. Matt walked me through it: fill up the big kitchen sinks with ice water; take the live crabs from the bushels, and, using tongs or rubber gloves, plunge them into the ice water, thereby knocking them out. Then, when he gave me the signal, I had to put the iced, comatose crabs into a large bus pan and take them over to the big pot that Matt would drop them in. After they were cooked, Matt would give them to me to get ready for the customers.

It seemed like a lot of hubbub for the result: live crabs became cooked crabs. Why not cut out the middleman and just take the crabs directly from the bushel and drop them into the pot, I asked Matt. Was the ice water coma a sort of anesthesia, an attempt at being humane?

"Nah," Matt said, smoking the same cigarette. "If you drop 'em in when they're alive, they flip the fuck out, and when they cook, their claws and legs and all the shit drops off. Then you're pulling just crab bodies out of the pot, and all the limbs sink to the bottom. You ice 'em up so they don't flip the fuck out and they stay whole. People want the whole crab on their plate, not the crab and a pile of claws and legs and shit." Matt walked back into his kitchen.

"Fuckin' anesthesia," he laughed. "They're fuckin' crabs!"

I took the lid off the first bushel. A single crab scooted over his brothers and sisters, escaped, and fell to the floor, making a loud cracking noise. "And don't let them drop!" Matt yelled from the back of the kitchen. "We don't want no broken shells!"

I quickly put the lid back on and regarded the wounded escapee, crawling around the floor. I used tongs to pick the crab up and place him delicately in the empty sink. The clickety-clack of its claws and legs against the sink echoed through the kitchen, over the Dire Straits song playing on the small radio above my head.

I put the escaped crab back with its brothers and sisters. Then I filled the sinks with ice water and reopened the lid of the first bushel. The crabs were indeed scary. They were bigger than I'd imagined, they stank of the ocean, and there were dozens and dozens of them, all claws and creepy little legs and antennae, squirming and crawling. And there was the noise. Thousands of little clicks, clacks, and scratches, clicks, clacks, and scratches, clicks, clacks, and scratches.

I was overcome not with fear but with a sense of the responsibility of my mission. I was the last person these crabs would see alive. If they could, in fact, see. It looked like they had two beads or whatnot for eyes, so I imagined they had some eyesight. Again, not a big crab person.

What I had to do was important. I was going to kill a few hundred living things. Though I was neither vegetarian nor hippie—the very sight of cows made my mouth water—the least I could do was make their journey from bushel to sink to pot as comfortable and quick as possible. So I pushed my fear aside, put on the gloves, grabbed the tongs, and started gingerly loading the crabs into the deep, icy water in the sinks. Seconds after being immersed, they were "asleep."

"That's it, Jay, you got it," Matt yelled from the kitchen. "Drop 'em on in. Keep going."

Over time, I got better and more efficient at being the crab man. I got braver, too. I ditched the tongs and would use only the rubber gloves. I eventually lost those as well, and would grab the crabs quickly with my bare hands, moving back and forth from bushel to sink, bushel to sink. The few times I got pinched did not hurt so much. And these were war wounds, battle scars, proof of a tough job done well.

About a month after taking on the new job, I was flinging the crabs into the sink sans tongs and gloves when Mike walked into the kitchen.

"Look at him, Matt. He's a natural. No fear at all!"

In a few months, I was going to start my sophomore year. I was still afraid of bugs and people with dreadlocks. But, aside from David, I'd spent more time in a bar than any kid I knew. And I was a fearless crab man.

philadelphia's finest

Every single night growing up, my family would eat our "treats." As the oldest child, I was charged with walking to one of the corner stores, Hennessey's, which was also a lo-fi ice cream shop. I say "lo-fi" because there were no colorful banners with cartoon characters or ice cream cakes available for purchase, or sundaes that came with fancy toppings like shaved chocolate truffles or crumbled butterscotch brownies. Along with your gallon of milk and your paper and your roll of toilet paper, you could get a cone of ice cream or a milkshake or, if they had enough of the ingredients around, maybe a banana split.

We all had our usual orders, which included a main choice and a back-up, should the preferred ice cream not be available. For my dad, it was a double dish of banana (or strawberry). For my mom, a cone of her beloved chocolate marshmallow (or vanilla fudge). My brother and I were the same: a cone of cookies 'n' cream or vanilla, though sometimes I changed it up, making a game-time decision once I got to Hennessey's. My sister, seven years younger than me, would carve out a spoonful or two of my mom's treat.

This was not for special occasions or an every Saturday night thing, but, as mentioned, *every single night*. When we moved to my grandmother's house a few blocks away while my parents settled the whole divorce thing, we continued the ice cream tradition, going to a place called Bell's instead of Hennessey's. From the time I was five until just about the start of high school, I ate an ice cream cone right before bed every night. And those motherfucking cones were not small.

So the treats were strike one. Strike two was that I did not have a vegetable that was not a potato, corn, or prefixed by "creamed" or "cream of" until college. A salad was something that came with dinner when you went out to eat and which you pushed to the side after you picked the Russian dressing–doused croutons from it. I'm not going to play the "poor urban white" card and say that fresh produce was not available to us. It's not like there were multiple local organic farmers markets, but, Jesus Christ, we could have had some celery or a goddamn tomato every once in a while. I've since tried to get more vegetables into my diet, but it has been difficult. As an adult, I dated a girl whose family owned a farm, and when we went to visit her family, they would hold up a vegetable and make me guess what it was. Then they'd break up laughing when what I thought was a carrot was really a radish. However, I'm happy to report that my family is broadening its horizons in the veggie department somewhat; at our most recent Thanksgiving, in addition to the mashed potatoes, sweet potatoes, whole corn, creamed corn, broccoli with cheese sauce (from a microwavable bag), and creamed spinach (also from a microwavable bag), we had green beans. They were in a casserole, loaded with cheese and onion rings, but it's a start.

What we did eat on a regular basis was what could best be described by one of my favorite adjectives, *rib-stickin'*. While my mom was a good cook, she was not known for the breadth of her offerings. Her meatloaf was astounding; a thick rivulet of cheese oozed through its middle. Her lasagna was very good, loaded with ricotta cheese and full of ground beef. Her chicken cutlet, which she could turn into chicken parm with some American cheese and Ragu spaghetti sauce, was out of this world. My dad didn't cook so much, but when he did, he made what we called "hash and noodles": a pound or two of ground beef cooked in a pan in its own juices, covered in spaghetti sauce and mixed with egg noodles. Hot dogs and baked beans was a meal that both dad and mom cooked often. These are not exactly light fare.

We did not eat seafood, aside from the occasional tuna fish sandwich that found its way into our lunch bags. My mom once offered us microwavable fish sticks, trying to pass them off as chicken fingers, but my brother and I, fat fucks that we were, knew what a goddamn chicken finger tasted like and saw right through her ruse. The Great Fish Stick Switch nearly caused a fissure in the home, and to this day, we still do not talk about it.

My mom's specialty was her desserts. My favorite was her Hawaiian wedding cake: two layers of yellow cake with vanilla pudding and pineapple chunks between them and Cool Whip frosting with coconut sprinkles on top. Her pumpkin roll was always a big hit: moist, spongy pumpkin cake with a cream cheese middle and powdered sugar. But her most famous dessert was her chocolate chip cake: chocolate chip cookies reinvented in cake form with a top layer dusted with sugar, brown sugar, and cinnamon.

So, no veggies and no seafood, but plenty of heavy meals and sweets. This was strike two. Strike three was that I grew up in Philadelphia.

My beloved hometown consistently ranks among the most obese cities in America in those studies that sometimes appear on CNN.com and subsequently get forwarded to me by my non-Philly friends with comments like "you must be so proud!!!" or "Philly: fat chops central!!!" Yes, our residents aren't known for their fitness, as are the people of Washington D.C. or San Francisco or other fit cities that I can't think of right now. But there's a reason for this: our local foods are so much better than theirs.

And so I present to you a brief overview of the foods that made me fat. These were staples of my diet growing up. Though I could have made healthier choices, I have no regrets. This stuff is delicious.

The Cheesesteak

When you think of Philadelphia and its foods, you think first and foremost of the cheesesteak. (Well, you should—I'm not exactly sure what's going on in your head.) Sometimes stereotypes are true. For example, studies have shown that four out of five black people are really good singers. The stereotype of the cheesesteak-loving Philadelphian is also true. I cannot understate the ubiquity of the cheesesteak in my life. If a week went by in which I didn't have a cheesesteak, it was because I'd spent seven days unconscious.

But while the cheesesteak is the food most often associated with Philadelphia, it is also the most often misunderstood and bastardized. To start, there are only four components to a Philly cheesesteak:

1. Bread
2. Meat
3. Cheese
4. Fried (grilled) onions, which are optional, but encouraged

That's it. If you see an "authentic Philly cheesesteak" as the special at your local Friday's or Chili's with these four ingredients plus peppers or mushrooms or whatever else, please ask to speak to the head chef. When he greets you, say hello, then rear back and punch him directly in his mouth. While he's probably just following orders from corporate headquarters and the "authentic" Philly cheesesteak is not his creation, great injustice calls for great reaction.

Opinions vary about which ingredient is the most important for the cheesesteak. Some will say that the bread must come from a famous Philadelphia bakery to be considered a true Philly cheesesteak (and so you have cheesesteak places in New York or Washington claiming their true authenticity because they have rolls shipped from Philly every morning). In my opinion, the texture of the roll is more important than the name of the bakery from which it comes. In this case, a proper cheesesteak requires a roll that is not hard or flaky or seeded, but rather one whose texture is soft and is slightly chewy. That it is *slightly* chewy is key. You need a roll with the density to withstand the grease of the beef and the cheese and the onions, but not so dense that it overpowers the other ingredients or tires out your jaw.

Others say the meat is the most important. Philadelphia itself is divided as to the presentation of the meat, which is usually rib eye or perhaps top round. The two camps are (what we will call) slabs and shredded. Some places offer their beef in three or four whole pieces, thinly cut and layered on the roll. Others will present it chopped or shredded, piled on the roll. There are benefits to each, but I am firmly in the shredded camp. With the slab style there is a need to tear each bite from the sandwich, whereas the shredded beef style allows for nice, simple bites (and quicker consumption).

There are three—and only three—types of cheese found on a real Philly cheesesteak: Cheez Whiz, provolone, or American. That's it. No Swiss or pepper jack or blue or anything else. Just Whiz, provolone, or American. John Kerry, during a visit to Philadelphia during his presidential campaign in 2004, caused quite a stir when he ordered his Philly cheesesteak with Swiss cheese. Swiss! Nothing says "I have no common touch" like ordering goddamn Swiss cheese on a cheesesteak. I bet if you ask him privately after he's had a few drinks why he lost the election, he'll point to that cheesesteak blunder.

I prefer the cheeses in the order in which I listed them. I understand that Cheez Whiz is neither technically nor legally actual cheese, but it is delicious nonetheless. Instead of viewing it as a "cheese product," I like to think of Cheez Whiz as advanced cheese. The good people at Kraft took a natural product and made it artificial, but just as good (if not better) than the real thing. Is that so wrong? For example, natural breasts are wonderful. So are fake breasts. We can argue all day about this (and I have the time to do so).

Provolone is a strong number two, and one that I do not shy away from. I never get American; this is the default cheese when getting a cheesesteak (or "steak and cheese") outside of Philadelphia. I choose Cheez Whiz and provolone not only because I prefer them, but because I have to take advantage of their limited availability.

The cheese and onions determine how the cheesesteak is ordered. "Wit'" and "wit'out" mean with or without onions. Add the type of cheese before this, and you have the basic construct for ordering a cheesesteak like a real Philadelphian: "Whiz wit'" or "provolone wit'out," for example (I'm a "Whiz wit'" man).

Finally, where is the best cheesesteak in Philly found? You may know of two competing cheesesteak places, Pat's and Geno's, who have a decades-old rivalry and sit cattycorner from each other in the heart of South Philly. On one side of the street you have Geno's, brightly lit with neon. On the other, you have Pat's, about as no-frills as you can get.

But as a native Philadelphian who grew up less than a mile from Pat's and Geno's, I can tell you that Pat's and Geno's are only good if a) you are a tourist and want to get your picture taken at a famous Philadelphia cheesesteak place or b) it is 3:30 in the morning and you are bombed and would eat your own hand if it was covered with Cheez Whiz (both shops are open 24/7).

When I'm in Philly, I'll patronize any number of local neighborhood cheesesteak joints that blow the best cheesesteak places in New York City, Chicago, and Los Angeles out of the water. But if you're asking my advice and you want to eat at one of the famous places (read: one likely to have been featured on the Food or Travel Channel), I would recommend Jim's on South Street or Tony Luke's in South Philly, just a few blocks away from where I grew up. Both those places will treat you right. Promise.

Congratulations! You can now converse about cheesesteaks like a real Philadelphian.

The Soft Pretzel

Though less popular than the cheesesteak, the soft pretzel is almost as dear to Philadelphians. According to Wikipedia, the average Philadelphian consumes twelve times as many pretzels as the national average.* Even if you figure that Philadelphians are two or three times fatter than the national average, that's still a lot more pretzels than everyone else.

The Philly soft pretzel has a different shape than your traditional soft pretzel. Unlike the more familiar bowed German-style soft pretzel, it is shaped like a very narrow 8, or an *H* with a closed top and bottom. The pretzel itself is lightly salted, dense,

---

* Yes, Wikipedia is the extent of the research I'm willing to put into this project. Get used to it, because you'll see it again.

doughy, and chewy. It is also rather versatile. While mustard is the topping of choice, I've had my fair share of pizza pretzels for lunch. The warm sauce and the melted cheese on top of the soft pretzel is so magnificent that after eating one, you might ask yourself why it hasn't taken over the world.

Man is no closer to God than when he eats a warm soft pretzel fresh out of the oven. If you have this opportunity, take it and do not look back.

The Hoagie

I've eaten a lot of hoagies, subs, and heroes in my life, so I have a lot of empirical evidence for this one. I say the following with confidence: there is nothing major that differentiates what Philadelphians call the "hoagie" from what others call a "sub" or a "hero." It's a sandwich, usually with lunchmeat, cheese, fixin's, and maybe some oil or vinegar or mayo, on a long roll. (However, I would venture to guess that Philadelphia has more men nicknamed "Hoagie" than any other major city. So we have that going for us.)

There is one minor difference, however. Hoagies contain only lunchmeat. So, for example, there is no meatball hoagie or chicken salad hoagie, whereas a meatball sub or a chicken salad hero is common. You can get an Italian hoagie or a turkey hoagie or a ham-and-cheese hoagie, but the main ingredient must remain in the lunchmeat family.

Unlike the cheesesteak, with its simple list of ingredients, the hoagie is all over the place; with various meats, cheeses, fixin's, condiments, and types of rolls, the sky is the limit. And as long as you ask for a hoagie when you order it in Philly, we're cool.

Water Ice

Unlike the hoagie, there are distinctions between what Philadelphians call "wudder ice" and what the rest of the world calls Italian ices or snow cones. Italian ices are much more granular than water ice, while a snow cone is larger chips of ice with syrup poured over them. Water ice is much smoother in consistency, and juicier than either Italian ices or snow cones. Like Italian ices, the flavors in water ice are blended while being frozen. The result? A refreshing summer treat that is chockfull of sugar. Add some soft serve ice cream for what is known as a gelati. If this doesn't put you on the right track to diabetes, nothing will.

Creamed Chipped Beef

Creamed chipped beef (or CCB) is not unique to Philadelphia. It is known throughout the Mid-Atlantic region and gained notoriety in World War II, when it was served en masse to GIs and known as "shit on a shingle." It is generally considered a breakfast food and is served on top of white toast, English muffins, or biscuits. You may be familiar with its more popular hick cousin, biscuits 'n' gravy, which is generally available below the Mason-Dixon line every two or three miles.

True, creamed chipped beef has a bit of a damaged reputation—it's hard to overcome a nickname like "shit on a shingle." But its four main ingredients are milk, butter, corn starch, and thin shreds of a dried, salted beef that is something like bologna but even lower quality. I ask you, friends: what is wrong with any of those ingredients? The milk and butter make up most of the mixture, and corn starch stiffens it to create an almost gelatinous texture. And there's little pieces of salty beef sprinkled throughout. Sure, the beef is not high-quality and by itself might taste really bad; it's sold in some supermarkets in jars and smells kinda like cat food. But you could cover shards of glass in a milk-butter–corn starch combination and I'd eat the shit out of them.

Food of kings it is not. But CCB is, after the cheesesteak, my favorite homemade food. My dad has never been a big fan. Every week while growing up his family had a "shit on a shingle" night when he and his brothers and sisters would take their pieces of toast and plates and line up to be served by my grandmother, cafeteria-style. But I have no such childhood memories, and eat CCB every chance I get (which is not very often). Over the years, CCB has become increasingly difficult to find, as chain restaurants that you might expect to serve such an artery-clogging, tasteless, and cheap-to-make breakfast food are taking it off their menus.

Is it a food worth traveling for? No. But if you are not from the Mid-Atlantic area and see it on a menu, give it a try. I promise it'll taste much better than "shit on a shingle" sounds.

Scrapple

You know how Native Americans would use each part of the buffalo so as not to waste anything? Not only would they eat all of it, but they'd use the skin for clothing, the hair for rope, the hooves for glue, and the horns for cups or prosthetic noses or whatever? Well, if you read the ingredient list of scrapple, you'd think it was invented by Native Americans. Or maybe by a really savvy pork executive.

BUTCHER
(looking over a pile of pig detritus)
Well, we made the bacon and the ham and the sausage, and we got the cuts for the ribs and the pulled pork and the roasts, but we still got all this shit left over.

PORK EXECUTIVE
(a beat, thinking)
Fuck it—mash it all up, throw in some cornmeal and flour, put it in a loaf, and call it "scrapple." I'm sure somebody'll buy it.

BUTCHER
Done.

So what is in scrapple, exactly? According to Wikipedia:

[S]crapple is typically made of hog offal, such as the head, heart, liver, and other scraps, which are boiled with any bones attached (often the entire head), to make a broth.

Once cooked, bones and fat are discarded, the meat is reserved, and (dry) cornmeal is boiled in the broth to make a mush. The meat, finely minced, is returned to the pot and seasonings, typically sage, thyme, savory, black pepper, and others, are added. The mush is formed into loaves and allowed to cool thoroughly until set.

So . . . who's hungry?

But the thing is, it's delicious. It took me a little while to get used to it, because the ingredients—pork hearts, pork livers, pork skin—are listed in all their glory right there on the package you'll find in the grocery store. But grill it up all nice and crispy and put a little ketchup or mustard on it, and you'll have yourself a divine breakfast meat. Just don't think about it too much.

Pork Roll

Our second breakfast meat, pork roll, is very popular in the Philadelphia tristate area and is known by its brand name, Taylor Ham, in parts of northern New Jersey. Its closest relative is Canadian bacon—you know, the meat on the McDonald's Egg McMuffin—but pork roll is saltier and has a slightly tougher texture.

On summer days, when we were off from school, my dad would take my brother and me out to a late breakfast or early lunch. We wouldn't go very far—only to the food truck on the pier he worked on. When I say "food truck," it may conjure up images of hipsters dishing up lobster rolls or kimchi or Tex-Mex/Asian/French fusion out of brightly painted, specially designed automobiles. This was not the kind of food truck we ate at. First, I don't even think it was a truck. It's very likely that it was a small shack with a stove and a few tires out front. And there were no hip foods there. The word *menu* could have been replaced by "Foods that Will Substantially Cut into Your Life Span—And by the Way, We Do Not Have a License Here."

Each time we went, my dad and I would each get the slider—a pork roll, egg, and cheese sandwich—which we would coat with ketchup. Contrary to its name, this was not a small sandwich, but it was called a slider because it would slide right through you. It was best to consume this sandwich while wearing a diaper, if possible.

Whenever I have a pork roll, egg, and cheese sandwich now, I can't help but think of that food truck on the pier on the Delaware River. Then I can't help running straight to the bathroom. The magic lives on.

Tastykake and Wawa

Tastykake is a Philadelphia-area bakery similar to Hostess or Entenmann's. Wawa is a Mid-Atlantic-based convenience store similar to 7-Eleven. I've lumped them together because they're not really foods, per se. Both Tastykake and Wawa were a daily presence in my life. A Tastykake was a common breakfast in high school, while Wawa was a multiple-visits-per-week kinda place, for everything from groceries to quick sandwiches.

There is not much to add, except that both Tastykake and Wawa are so far superior to their competitors that those competitors should be ashamed of themselves. Knowing only Entenmann's or Hostess or 7-Eleven and then experiencing Tastykake or Wawa would be like being born and raised in an Amish community and then getting a summer job at the pool at the Hard Rock in Vegas. Nothing wrong with the Amish way of life, but you should know that there are many decadent (and delicious) things out there. My affection for them is so great that if I ever win the lottery, I will build a Wawa and a Tastykake factory into my home (right after the construction of the grotto filled with statues of the Spice Girls is finished). So I guess I'm saying that they have my full endorsement.

So Let Me Introduce You to the One and Only Billy Shears

Though we were the same age, my buddy Vic was like a big brother to me. He was a lot larger than I was, so he looked the "big" part. But he had an older brother, Billy, and it was through Billy that Vic knew shit.

When hanging out at the Park, the local playground at which we congregated on a nightly basis, Vic knew and spoke to all the olderheads—his brother's friends a few years older than us—whom we alternately feared and admired. Both of us were a long (long) time from getting laid, but the first drink I ever had was at Vic's house, a shot of warm gin at a party after our eighth-grade graduation dance. I threw up immediately, and then we shaved Vic's head. Killer party.

Vic, through Billy, was the first of our friends to adopt classic rock. Hip-hop, rap, and R & B was the default music for most of us, presented to us by Power 99 FM or Q102, the two "urban" radio stations in Philly (though the latter was more pop-oriented than the former). We listened to this music because we enjoyed it, but also because it was cool. It was modern, it was edgy, and it was not what our parents listened to. (To this last point, no, it was not quite like the World War II generation who grew up on Fred Astaire spawning hippie children who moved to San Francisco so they could do drugs and sleep with three-fifths of Jefferson Airplane. But this was a white working-class neighborhood in which most of the kids would list EPMD, Pete Rock & CL Smooth, and A Tribe Called Quest as their favorite musicians.)

Vic and I would argue about the merits of classic rock vs. hip-hop. He said that his music, bands like Led Zeppelin and the Beatles and the Rolling Stones, had stood the test of time and would be around forever, unlike the passing-fad crap that I listened to. I countered that if we kept listening to oldies, there would be no oldies in the future. For example, did our parents listen to music from the 40s and 50s when they were kids? No, they listened to the modern stuff. The bands Vic liked were dinosaurs, for old people and parents.

Vic's family had a house "up the mountains," Philly-ese for the Poconos. I was invited to the house one winter weekend with Vic and his parents, his sister, and her friend. I had never been to Vic's place and was happy to go. Of the two vacation options for my family and friends, going "downnashore" to North Wildwood, New Jersey, or "updamountains" to the Poconos, I preferred the latter to the former. I loved going to the Jersey Shore, but the activities up in the mountains—mostly walking

around in the woods with BB guns shooting at trees and drinking hot chocolate—were slightly more my speed than those of
downnashore, where we sat at the beach or pool, getting sun-
burned.

No sooner had we reached the mountain house than it
started snowing, and it did not stop until a day and a half later.
We were snowed in with nothing to do, aside from playing video
games and talking. Soon, Vic and I picked up our eternal debate
about his music vs. my music. Exasperated and stricken with
cabin fever, Vic said, "Look, just do me one favor? I want you to
listen to a couple of songs off this CD while I get in the shower.
Just the first couple of songs. If you don't like it, you win and it's
over—your music is better than mine. But you have to give it a
listen." Fine, I said, and he threw me the CD and went down the
hall to the bathroom.

It was a Beatles CD. Like I didn't know the Beatles already.
Who didn't know the Beatles? They were on a regular loop on
the oldies station: "Help," "Ticket to Ride," "Twist and Shout,"
"I Saw Her Standing There," "She Loves You," "I Want to Hold
Your Hand," "Can't Buy Me Love," "A Hard Day's Night." I
knew, like, every Beatles song already. The cover was pretty
cool, with the Beatles dressed up like rainbow generals with a
bunch of cardboard cutouts behind them. But I was sure this
CD wouldn't show me anything I didn't already know. As he
walked out of the room, Vic put the CD on his stereo.

Before long, I was confused. This was not the Beatles. No
way. Not the Beatles that I knew. This was like a whole different band, because there was no way that the same guys who
wrote "Eight Days a Week" could write this "Lucy in the Sky

with Diamonds" song. Vic came back into the room a few songs later and I waved him off. I sat there listening until the album was over. It was the weirdest and best album I'd ever heard. So I put it on again.

On that snowy weekend in the Poconos, Vic and I listened to *Sgt. Pepper's Lonely Hearts Club Band* about four hundred times while we played video games. Vic would sometimes break it up by putting on *Revolver* or *The White Album*, but I wasn't through with *Sgt. Pepper.* I needed to hear it again and again (and again).

I know that I'm not breaking new ground here, that there are thousands of stories of young people having musical epiphanies because of *Sgt. Pepper.* I'm guilty, and I wish—for your sake, dear reader—that it was something more original, like Jethro Tull's *A Passion Play* or Yes's *Tales from Topographic Oceans* that turned me on to rock music for good.

Alas, *Sgt. Pepper* was my gateway drug. From then on, I worked my way through the rest of the Beatles catalog. On Vic's suggestion, I started with *Rubber Soul*, which was the first of the "adult" (Vic's word) albums. Then I made my way chronologically all the way to *Let It Be* before getting back to the bubble gum Beatles stuff, their earliest works with which I was already familiar.

I've never dedicated myself to something so completely and diligently as I did to the Beatles in the weeks after the Poconos trip with Vic's family (and yes, I'm including masturbation in this assessment). I don't think I listened to anything else for the next few months, instead committing myself to learning every nook and cranny in every song on every album. Every drum roll, every harmony, every line, every lyric. I read, too, starting with CD liner notes and then moving on to book after book about the band.

The Beatles crept in to every facet of my life. A few weeks after the Poconos trip, Ms. Flynn asked us to write a paper in English class on a poem or song or short story that included a minimum of five of the things we talked about when we analyzed works of literature, like simile/metaphor, use of imagery, rhyme, alliteration, and the like. I chose "Lovely Rita," from *Sgt. Pepper*, which really pulled off alliteration without sounding corny ("sitting on a sofa with a sister or two") and made use of all the other stuff, too. Freshman year of high school, our history assignment was to write a five-page paper on whatever we wanted, as long as it was "historical" and approved by the teacher, a six-foot-six Jesuit priest in his late fifties named Father Taylor. My paper was on the Beatles's alleged drug use and its effect on their music, starting with "Tomorrow Never Knows" on *Revolver*, through *Sgt. Pepper* (with "Lucy in the Sky with Diamonds" as the primary example), and on to *Magical Mystery Tour*, *The White Album*, and *Abbey Road*. I don't recall what grade I got, but it didn't matter. I would have written that paper in my spare time and fifty times as long.

My Beatlemania extended beyond schoolwork. Inspired by songs like "Helter Skelter" and "I Feel Fine," I got an electric guitar. It was a cheap model from Kmart, a black guitar with a built-in speaker powered by two 9-volt batteries. I wanted a Rickenbacker, like the one John played during the early years, but the ninety-nine-dollar, 9-volt-powered guitar already stretched my budget, so the Rickenbacker would have to wait for another time. Along with the guitar came an instruction booklet, covering the basics and some chords. And the four of us—me, the guitar, the booklet, and my Discman—spent hour after hour in my bedroom, playing "Revolution" over and over again, trying to get the right rhythm down.

I'm not sure who took this picture, but I am sure it wasn't my girlfriend. She was busy, um, living in Canada.

Finally, and most disturbingly, my appearance started changing. I'm not talking about normal growing-up stuff, like getting acne or growing bad facial hair. Though Paul was my favorite Beatle—his vocal range was more akin to mine, and he wrote the pop songs I enjoyed—I tried to make myself look like John Lennon. Well, John Lennon if he had let himself go and moved into a Pizza Hut for four years. First, I traded my uncool glasses for the little round ones that John wore. Despite the fact that these glasses were too small for my face and made my cheeks look like they were growing around the glasses, I loved them (even though I had to pry them from my head before bed). Then I started growing my hair long. Of all the things I did or did not do as a teenager, it is the long-hair phase that is chief among my regrets. Parted right down the middle, my long hair would naturally flip or curl upward, just below my ears. The flip looked intentional, like I spent the morning drunk in front of a mirror with a curling iron. I don't remember if I thought my long hair looked cool or if I was trying to be nonconformist. But when I look at the pictures now, the only reasonable explanation for that hair is that though I was pretty sure I wasn't going to have sex with a girl, I wanted to be absolutely, 100 percent sure that I was not going to have sex with a girl. "Pretty sure" was not good enough. I needed to keep my virginity intact.

But back then, I remained blissfully ignorant of my appearance and unaware of how it might affect my standing with the ladies. The only thing that mattered to me was the music. The Beatles were my idols. If I wanted to spend half my time in high school looking like Fat John Lennon, whatever. It was all about the music, man.

the six most influential songs of my teen years

As much as I loved them, it wasn't all Beatles, all the time. Here's a sampling of a few of the most influential songs of my teen years.

"Buddy Holly" Weezer

Nirvana and Pearl Jam were good bands. If my kids, whom I hope will love music as much as I do, ask me what it was like when Nirvana and Pearl Jam burst onto the scene, I'll tell them all about the flannel and Seattle and how everybody learned how to play "Smells Like Teen Spirit" on their guitar. I'll say, "Yep. It was really something." And then I'll look back at the television and go quiet for a while because there's a baseball game on that I'm trying to watch and enough with the questions already.

I liked both Nirvana and Pearl Jam because, as a twelve-

year-old boy when *Nevermind* and *Ten* were released, I was basically required to. But overall, there was a little too much angst
in those albums for my taste. All the anger and screaming and
unintelligible lyrics wasn't really my thing. I mean, I knew it
rained a lot in Seattle, but—get over it, guys.

This doesn't mean that I eschewed all modern rock. In my
opinion, there was one band that stood head and shoulders
above the rest, pioneers of geek rock who could deliver both the
crunchy power chords and the lyrics about heartbreak in the
same song: Weezer.

Weezer had angst, but they also had cleverness. Weezer had
distorted guitars, but they also had harmonies. Weezer could
rock, but they also could be quirky and smart and nerdy. I
wanted to rock, and I was quirky and smart and nerdy. But I list
this song here not only because I loved the band.

Dennis and I as the Pound Patrol, with two very little people (sister
Megan, cousin Marisa). I was a big fan of UNLV. And I mean *big*.

My brother and I were both fat as kids. Our collective chubbiness made us the object of derision for many a neighborhood asshole. To wit, one of the neighborhood nicknames for my brother and me was "The Pound Patrol," meaning Dennis and I went around (on patrol) searching for hoagies and cakes and whole milk. Kids can be very cruel.

But of all the barbs slung at Dennis and me over the years, none stung as much as the little ditty made up by my "friend" Stephanie, a girl in the grade below and a member of our crew, which was sung to the tune of the popular Weezer song "Buddy Holly." Imagine the song with me, if you will, and feel free to sing along to Stephanie's revised chorus:

> *Woo-ee-oo, I look just like Jason Mulgrew.*
> *Oh-oh, and you're Dennis Mulgrew, too.*
> *We don't care what they say about us anyway.*
> *We don't care about fat.*

Woo-ee-oo. We don't care about fat.

The song took the neighborhood by storm, and soon Dennis and I were greeted with "woo-ee-oo" at every turn. It is a small wonder that neither Dennis nor I is a serial killer targeting clever, somewhat musically inclined fourteen-year-old blond girls who tease fat kids.

But it all worked out in the end. I'm not sure what became of Stephanie—I lost track of her over the years. But I'd say that best-case scenario for her is a cheap motel, various overweight men, and a lot of oral sex in exchange for meatball subs. Dennis is a graduate of a top law school and is all sorts of boss. And, perhaps because of Stephanie's little song and the resulting torment he endured, Dennis is now such a health and fitness freak that even the veins in his biceps have abs.

And me . . . well, I ain't doing too shabby. Yes, I'm still a bit chubby, but I'm a writer, so that's pretty sweet. A writer who may be softly crying as he writes this, but a writer nonetheless.

"Linger" The Cranberries

Though the "Buddy Holly/We Don't Care about Fat" remix was a big hit in the neighborhood, artistically it had nothing on the parody of the Cranberries' "Linger," which became more of an indie or cult hit. Let me explain.

In eighth grade, our friend Josh developed a crush on a girl named Christina. My friends and I, Josh included, did not know much about Christina, who was cool and mysterious because she didn't go to our junior high and also because she smoked Marlboro Reds.

But Josh was always one for a challenge, and one day he got up his courage and asked Christina out on a date. They went to the local eatery, the Oregon Diner, for a dinner of French onion soup, chicken fingers, and broccoli puffs. After the dinner, he fingered her.

(!!!)

As Josh was the first among our group of friends to confirm that this strange thing called a vagina actually existed, this rocked our social circle. Remember, though we grew up in the city, we were Catholic school kids with very little hands-on knowledge of, or experience with, sex. And here was Josh, one of our own, fingering (!) a girl's vagina (!!!).

Soon our whole clique of friends knew about Josh's finger-ing of Christina. And someone (I'm not sure who—I wish I could take credit) made up a little parody of the Cranberries' "Linger," which was very popular at the time:

> *But I'm in so dee-eep.*
> *You know I'm such a fool for you.*
> *You even took me out to dinner, uh-huh.*
> *Did you have to use your finger?*
> *Did you have to*
> *Did you have to*
> *Did you have to use your finger?*

Because she didn't run with our group and didn't go to the same school as we did, I'm not sure if Christina ever heard our song. She hadn't been around much before the Great Fingering Incident of 1993, and she wasn't around much afterward, either.

I haven't seen Christina in years; she would now be in her thirties. But every time I hear those first few notes of "Linger" and Dolores O'Riordan humming along, I always—and will al-ways—be reminded of Christina, a true trailblazer in the world of digital penetration.

"Stay" Jodeci

Among the things I knew very little about as a teenager, two at the top of the list were sex and black people. This song ex-plained them both.

"Stay" is about post-breakup sex: I screwed up, and you left. I am sad, but you're here now, so why don't you just stay, and we can make love? What do you think?

We don't know what the female's response is, but at the end of this song, *I* certainly feel like staying and having sex, so we can only assume that's what happened.

This song was one of the mainstays of my favorite radio program, *The Quiet Storm*, which featured "smooth, slow jams" from the late 70s to the then-present. Even after I converted from hip-hop to rock 'n' roll, *The Quiet Storm* was my secret pleasure because I could not find any rock equivalent to the sexy, R & B sound. White people had power ballads—songs about I remember you and every rose has its thorn and you don't know what you got till it's gone. Black people had songs that said, baby, I miss you, stay with me and let's get down, real slow-like, just for one night. When it came down to a soundtrack for lovin', white songs did not even come close. And though I wasn't having any lovin', it was important to know which music I'd play should that opportunity present itself: "Stay" by Jodeci, one hundred times in a row. (Although, let's be honest, just the first chorus would probably suffice.)

"Crash into Me" The Dave Matthews Band

Just because a song is included among the most influential of my teen years does not mean that I liked that song. Remember, Adolf Hitler was once *Time*'s Man of the Year. Being influential is not always a great thing.

I hate this song. And I hate this band. I have nothing against Dave Matthews personally. He and his bandmates seem like lovely guys. And I get the whole scene: the dancing and the fun and the really fit violin guy and the "ants marching" and all that crap. It's great, it really is. Good for them.

Dave Matthews had already endeared himself to white people everywhere with *Under the Table and Dreaming*, which, in my circle of high school friends alone, sold approximately forty thousand copies. The girls were especially under the spell of Dave, a safe, sensitive, but fun guy. Following "What Would You Say" and "Ants Marching" with the love song "Satellite" only made him more desirable to this demographic.

Say what you will about "Satellite": at least it was difficult for high school guitarists to play and sing. When guitars were suddenly broken out at high school parties ("What? Me? Play? I couldn't. . . . Oh, alright."), a few would try, knowing that a Dave Matthews love song was the easiest, pre–Miller Lite way into a girl's heart (and later, her pants), but few had the talent to pull off the song.

The same could not be said for "Crash into Me." In many ways, it was ideal for the high school guitarist. The riff (and chorus), which is instantly and immediately recognizable, is not so easy that just anyone can play it, but is not difficult to master for anyone who is a capable guitar player. It's also an easy song to sing—even Dave speak-sings his way through most of the song. (If you could nail the high "Crash into me" during the outro, major bonus points for you.)

By itself, the song is nice. I can appreciate it. But for me, my dislike of this band and this song comes down to one thing: sex. Every generation has a song that becomes the soundtrack for lost virginities. This was my generation's. Dave Matthews wrote this song, and hundreds of thousands of white high school guitarists played their way into sex with it. And every time I hear it, I'm reminded that I was not one of them. So fuck this song.

"The Wind Cries Mary"
The Jimi Hendrix Experience

When I'd saved up enough money, I decided it was time to get myself a real electric guitar. I didn't have a specific model in mind, because I was working under the constraints of a modest budget. But I figured that anything would be an upgrade over my current 9-volt-powered "ax."

I stopped in a local guitar shop, just off of South Street, and was immediately met by a sales rep. He was in his late thirties or early forties, was skinny and had tattoos, and almost certainly had a criminal record (but for nothing major—maybe just possession or loitering). He asked me what I was looking for and I explained my situation.

The guitar salesman walked over to a wall of guitars and pulled down an off-white Fender Stratocaster. It looked just like the one Jimi Hendrix played at Woodstock. It had caught my eye when I walked into the store, but I assumed it was out of my price range.

As he started to plug it in, he asked me, "You know Jimi Hendrix, right?" When I responded in the affirmative, he said, "Then you probably know this one." He checked the volume on the amp, settled the guitar onto his shoulder, and played the first three chords of "The Wind Cries Mary."

At that point, he could have stopped and said, "The guitar costs fifteen thousand dollars. And to get it, you'll also need to rob that clothing store over there across the street for me." Instead, he played through the song as he spoke. "It's not American-made, but as you can see, it's got a real nice reverb sound. And it's a very good-looking guitar, and in your price range." At least, this is what I think he said—I was too busy focusing on the most beautiful noises coming out of an electric guitar I'd ever heard.

After the song was through, he asked if I wanted to try it. I did. But not after watching him play it. Instead, he gave me the guitar and I held it for a moment or two before telling the salesman that yes, I wanted it. No need to play it. I wanted it.

I took the guitar home that day and plugged it into the amp I'd borrowed from a friend and played "The Wind Cries Mary" (conservatively) two thousand times. I knew the song and had played it often on my crappy 9-volt guitar. But playing it on that off-white Strat brought me closer to Rock Godness than I'd ever felt before. That the guitar was not made in America like a real Strat and was perhaps made in Papua New Guinea didn't matter to me at all. For the first time, I felt like a real musician. A real, *cool* musician.

"How Could You Want Him (When You Know You Could Have Me)?" The Spin Doctors

"How Could You Want Him (When You Know You Could Have Me)?" could be the title of the romantic history of my high school years. My only edit would be to include a subtitle: "Loneliness, Lust, and Other Misguided Thoughts of a Teenager Who, My God, Would Do Anything for a Goddamn Girlfriend Already." I know it's too long, and I'm not married to it. Suggestions are welcome.

In this song, the protagonist expresses disbelief that the woman he feels affection for—and who seems to feel some affection for him—chooses to be with someone else. He frames this rejection within poetic, quasi-erudite lyrics, making references to *Hamlet*'s Guildenstern and Rosencrantz, seraphim and St. Christopher.

It's not quite right to say that I loved this song as a teenager; I *was* this song as a teenager. It's like my fifteen-year-old self wrote this song after a particularly painful house party at which my current crushee spent an hour in the yard making out with the captain of the basketball team. So this was my anthem, my solace whenever I was feeling blue. Just me, home from the party, alone in my bedroom with my headphones on, listening to this song, unsure of why I wasn't scoring chicks (as I ate a piece of cake), telling myself that I was a good catch (as I twirled my long, greasy hair), knowing that sometime soon, no matter what, I'd meet my soul mate (as I picked chunks of said cake out of my clear braces). Yes, someday soon I would find love (as I took off the sweatpants I wore to the party). The Spin Doctors understood where I was coming from. And so would my soul mate (as I finished the last of my chocolate milk).

lazy rider

By the end of my sophomore year, it became apparent that I was living a double life. Not in a mild-mannered-student-by-day/renowned-orgy-master-by-night sort of way. But socially.

Despite my misgivings about attending a high school at which the majority of the student body was from the suburbs or the nicer parts of Philadelphia and had eaten seafood before, I had little trouble fitting in at the Prep. When I first started, maybe I told a white lie here and there to fit in at a school where so many students came from well-heeled families—lies like my mom had sung back-up on the song "Ghostbusters," or that my dad roomed with Mike Schmidt for seven weeks in the winter of 1973. But it turned out that such fabrications were unnecessary. Indeed, there was not much difference between me, a nerdy kid from South Philly, and my buddy Kyle, a nerdy kid from Gloucester City, New Jersey, who, despite being only slightly tanner than a sheet of loose leaf and having the build of a walking diving board, was so well-versed in modern hip-hop that in some academic circles he was considered an expert on the Wu-

Tang Clan's seminal *Enter the Wu-Tang (36 Chambers)* album.
Nor was there much difference between me and my buddy Paul,
a nerdy kid from Havertown, a nice suburb of Philly, who, at
six-foot-six, could dunk a basketball but dressed like Robert
Smith from the Cure and worshipped Rush Limbaugh. We
were awkward teens, geeks throwing ourselves headlong into
our newfound passions, whether they revolved around the Wu,
conservative radio talk show hosts, or, in my case, the teachings
of Lennon-McCartney.

I must pause for a moment so that I can properly define the
word *nerd*, as I have been using it liberally. The term *nerd* immediately conjures up what I will call the *classic* nerd: someone
with thick glasses, acne, a pocket protector, and no social skills;
the guy who gets picked on by the bully who makes him do his
math homework. Meek and weak, there is nothing redeeming
about the classic nerd—not even his mighty intellect. My Prep
friends and I were not this type of nerd.

Then we have the *modern* nerd, far and away different from
the classic nerd. The modern nerd is cool. Modern nerds invent
things like Facebook and Twitter. Modern nerds form bands
like Arcade Fire and Vampire Weekend. Modern nerds make
lots of money and fuck your girlfriend in the bathroom of the
club when she and her friends have girls' night out and you and
your buddies stay in to watch a college football bowl game. My
Prep friends and I were not this type of nerd either.

We were somewhere in between the classic nerd and the
modern nerd. Part of the beauty of the Prep was that it fos-
tered an environment in which smart kids could bond because,

for the first time, being smart was cool. This was no nerd farm where students were sent off into basement labs to work on experiments or translate Cicero ten hours a day. The Prep's message was more akin to "you're here, you're smart, you actually might be cool." Coming of age in an environment that was supportive of nerdiness was far better than doing so in, for example, a large public high school that may or may not have existed solely to perpetuate stereotypes (jocks, nerds, cheerleaders, those weird kids that dress like vampires, abused kids, etc.), and it gave us confidence and allowed us geeks to flourish. At the Prep, we had good athletic teams *and* killer debate and chess teams. Our mixers, dances held a few times a year limited to Prep guys but open to girls from any high school, were always sold out and had a healthy girl-to-guy ratio—and we *still* had a higher proportion of virgins than any high school in the greater Philadelphia area.

So as you continue to read about nerds, please keep this in mind. We did not wear pocket protectors, but we did not go to NBA games with Justin Timberlake. We were something somewhere in the middle.

I still hung out often with my buddies from the neighborhood. After working at Mick-Daniel's or on weekday evenings, I went to the Park to sit around, shoot the shit, and hope for a fight to break out on the basketball court. But during the week, I'd stay after school and hang out with my Prep friends, to sit around, shoot the shit, and hope for a fight to break out on the basketball court. Thus the double life: I wanted to keep one foot in the neighborhood because I loved my friends and had known them forever, but at the same time I really liked the people I'd met in high school. And if I wanted to keep hanging out with my high school peeps, it became clear that I'd need to tackle that rite of passage in every teen's life: getting a driver's license.

While most teens look at a driver's license as a major step toward freedom and independence from the shackles of their parents' rule, as far as I was concerned . . . meh. Yes, having a license would allow me to travel of my own volition, which would be nice. But then I'd actually have to *drive*.

Most of my friends from high school lived in the suburbs west and northwest of the city. Under the current system, I got out to the burbs because my friends Kyle and Flem, who had already gotten their licenses and lived in New Jersey, would drive me. My house was practically on the way out to the suburbs, so they'd pick me up when we'd go out for a night of ordering a pizza and watching a movie in a friend's basement and hoping to God that Renée Garrity was wearing a shirt that showed some cleavage.

But Kyle and Flem wouldn't drive my ass around forever. When my time came, it was expected that I'd get my license and then join the rotation. I had very, very little interest in this. I liked things the way they were: Kyle or Flem would pick me up, and I'd sit in the back seat and make music suggestions or just look out the window while one of them drove and the other tried to navigate (this was before GPS, and we were fifteen- and sixteen-year-old kids driving around in unfamiliar areas) while I said things like "I really don't think this looks like the right way" and "Can we, like, stop for a soda or something?" and "No, but seriously, I don't think this looks like the right way." Why change a good thing?

Furthermore, in the other part of my life, I really didn't need a car. In the neighborhood, I could walk everywhere—to friends' houses, food joints, music stores, and South Street, Philly's counterculture center, with all its stores and people-watching. Growing up and living in the city meant that everything I needed to survive and thrive was no more than a walk away.

Then there was learning how to drive a car. As someone who couldn't ride a bike until he was nine (by that point, it was either learn or give up, since there were no training wheels strong enough to hold me) or swim until he was eleven (though I could wade without drowning from a young age), I was not looking forward to the process of learning how to drive. Instead of falling off your bike or having to doggie-paddle furiously to the side of the pool, if you fuck up when you're driving a car, you die in a fiery crunch of tons of metal.

Yet still, I am American, and in America, young boys are conditioned to want to get behind the wheel of an automobile and drive off, away from the restrictions of their parents and their town, free to be the person they want to be, to conquer the open road—all that crap. So a few months before my sixteenth birthday, as soon as I was eligible, I went to the DMV testing station in Southwest Philly to get my learner's permit. It was a dank, hopeless place, filled with parents and pimply-faced teens nervous about taking the permit test. I had been studying the booklet, but I was a bit nervous, too. I could not have been less interested in the rules and regulations of the road in Pennsylvania, so the information didn't stick easily. My name was called, and I was seated before a computer. After a few questions, I had my permit. Now I was ready to hit the road.

My Aunt Maureen and Uncle John surprised me with an early sixteenth birthday present: lessons with an instructor at a proper driving school. It was a relief, really. Not a small portion of my stress was related to who would teach me to drive. My mom was too high-strung; our only test drive had lasted all of one block, when she forced me to switch spots with her at the stop sign at the end of our street. My dad was at the opposite end of the spectrum and very laid back, but that didn't make the idea of learning to drive with him any better. He was a mechanic and devout car worshipper, and I couldn't imagine getting in the driver's seat while he sat in the passenger seat; it'd be like Johnny Unitas showing the proper way to throw a football to his son, whose only ambition in life was to work in clays.

The task of teaching me how to drive fell to Angelo the Driving Instructor, which is what his business card said. Angelo was less driving instructor than life coach. Whereas most kids learning to drive did so with a parent sitting next to them either yelling, crying, or pulling out clumps of their hair, Angelo was calm as could be (probably because he had a brake on his side of his specially designed driving-school car). His philosophy on driving was to instill confidence in the student: learn the basics, and you got it. "It's as easy as riding a bike," Angelo would say, unaware that if this was the case, we might have a problem.

Yet it worked. I found driving easier than riding a bike, because, unlike my parents and the various aunts and uncles who tried to teach me to ride a bike, Angelo was disappointment-proof. Stop in the middle of the street to wave a jaywalking pedestrian across the road? No biggie, but we shouldn't do that in the future. Let out a little yelp the first time on the highway?

Hey, that's fine. Proceed to drive on the highway in the right lane with the hazard lights on the entire time because there's no other way you can handle it? Of course you can handle highway driving—but we'll get 'em next time.

The Saturday after my sixteenth birthday, Angelo and I had an appointment at the same DMV at which I got my learner's permit, but this time, it was for the real deal. Angelo peppered me with his usual confidence boosters as we drove there.

"Think about it, Jase. Do you know how many people drive a car every day? Millions! And do you know how many idiots they give licenses to? Millions!" Angelo smacked the dashboard as he drove and delivered this inspirational speech. When he showed up that morning, I assumed *I* would drive to the DMV center; a sort of test before the test, it would require the most highway driving we'd done and to an area of the city with which I was unfamiliar. But when he picked me up, he motioned to the passenger seat, telling me that he'd drive so I could relax.

I knew the most difficult part of the whole driving exam would happen right away: the parallel parking test. Angelo explained that they did this first to weed out applicants: conduct the hardest part first, and if the applicant fails, you don't have to take him or her out on the road test.

Parking was the weakest part of my game, and if I so much as grazed any of the four orange cones, I would fail. But with Angelo's assurances and him sitting next to me, I thought I could do it. (It didn't hurt that the car I learned to park in was Angelo's small Toyota Corolla, which we were now pulling into the DMV center.)

Angelo got out and was replaced by a DMV employee. Without looking at me or up from his clipboard, he said, "Pull up and park between the cones." I tried not to think too much, remembering what Angelo had told me ("You know how many people parallel park every day all over the world? Millions!"). I pulled up to the designated spot, put the car in reverse, and backed up, turning the wheel right and then left to straighten out and gracefully sliding the car between the four orange cones. Just like that, before I could even think or worry about it, the hardest part of the test was now out of the way.

We were now onto the road test. You'd think some city planner would have realized that Colonial Avenue in Southwest Philadelphia was just about the worst place to put a DMV at which dozens of teenagers would be taking driving tests every day. Situated near a rail station, Colonial Avenue is a large thoroughfare perpetually clogged with traffic. It was almost as if the planning commissioners said to themselves, "Which area in Philly would you say is most like Hong Kong? We'd like to pack that area full of teenage drivers with less than ten hours' experience operating a car," and shortly thereafter, ground was broken on the Colonial Avenue DMV.

We made a quick right out of the DMV and found our way to the quieter streets behind the center. I was careful to remember my turn signals, to keep my hands at ten and two, to come to a full stop, wait three seconds while looking left and right, and proceed. After a few spins around, I made a left onto Colonial Avenue on our way back to the station.

Colonial Avenue itself has six lanes, three going in each direction. When I made the left onto Colonial Avenue, I had the luxury of turning at a light, so all I had to do was wait for a green turn arrow. But as we were approaching the center, I'd have to make a left across three lanes of oncoming traffic, without a turn lane or light to guide me, in order to get back into the station.

After pulling onto Colonial Avenue, the center in sight, I was feeling good—like I would soon be a licensed individual—when something unexpected occurred. An ambulance, just behind me, turned on its lights and siren.

The car kept moving, but my brain stopped. I didn't know what to do. I was in the leftmost lane, waiting to make that left back into the center. I knew that protocol was to pull over to the side of the road when an emergency vehicle got behind you, but I couldn't do that: my left was coming up and I had two lanes of traffic to my right.

I slowed down. I panicked. I slowed down some more. The DMV official asked what I was doing. The ambulance was behind me, all noise and red flashing lights, right on my ass. I prepared to make the left into the center and slowed to a near stop. The ambulance felt like it was crawling into the trunk of the car, like the medics were turning up the volume of the sirens. The DMV official screamed, "Move!"

I pushed on the gas pedal and jerked the wheel left. The car raced across the three southbound lanes. Oncoming traffic screeched to a halt. We pulled into the DMV center at about thirty-five miles per hour, so fast that the car cha-chugged as we drove up the entry ramp.

I slowed down as I pulled farther into the center as everyone—parents, students, DMV employees—stared at the car, wondering what idiot had just nearly caused a major accident. And there was Angelo, mouth agape, incredulous.

I brought the car to a stop. The DMV official ripped a sheet of paper off his clipboard, handed it to me, and said, "Better luck next time."

Angelo offered his best advice on the ride home, saying that it could have happened to anyone and that I had done good and that "millions!" of people fail their driving test on the first try. He dropped me off at home. And that was it. Our lessons were over, and I was on my own.

Yet I had a reason to feel, well, not terrible. I was bummed that I failed, sure. But an ambulance? Over the next few days, as the story was told to friends and family, it was seen as an unlucky break, an unfortunate but funny occurrence. After passing the parking and road tests, an ambulance! What bad luck! I laughed, too, and made an appointment the following week to finish this silliness once and for all.

This time, I went to the DMV with my mom. She wouldn't let me drive to the center, but she had been at ease being in the car with me when we had done a few loops around the neighborhood and made a few tries at parallel parking, most of which I aced without a problem. To make sure that, come test time, the parking would be as easy as possible, I was taking test #2 in the smallest car that anyone in the family owned: my grandmother's Geo Metro, which was about the size of a baby hippopotamus. You didn't so much drive this car as ride it. If you were tall enough and reclined the seat just enough, its hatchback could also serve as your sunroof.

At the DMV, it was the same familiar routine. Mom gets

out, DMV official gets in and tells me to pull ahead and park between the orange cones.

Viewing this whole test as more of a formality than anything else, I zoomed the car forward to the parallel parking area. I stopped, put the car in reverse, turned the wheel, stepped on the gas pedal—

And absolutely crushed the shit out of one of the orange cones.

It was so bad that it seemed like I had intentionally run over the cone. What other explanation was there? I was in a goddamn Geo Metro. You could practically drive that car through the door of a row house and park it neatly in the living room.

Maybe it was the cockiness, or maybe it was because I was jacked up, but it didn't matter: I was now 0-for-2. I'd failed my driver's license test. Twice.

The failure this time was no laughing matter and was much more depressing. Instead of being unlucky that an ambulance wound up behind me during test #1, perhaps I had been *lucky* that first time—lucky that I had so quickly and easily parallel parked the car and that I drove around the neighborhood without incident. Now it was possible that I'd never get my license, that I'd turn into *that* guy, the one who lives with his parents forever, takes the bus all over town, and buys a ton of porn videos but only when they go on sale. While I wasn't dying to get my license, this was getting bad.

I went into a funk. Following my life's motto, "If at first you don't succeed, immediately give up and focus on stuff you already know you're good at," I more or less quit driving and concentrated on, among other things, eating hot dogs and taking long naps.

Family, friends, and strangers all offered to take me out for a few spins, to drive to an empty parking lot and try out my parallel parking skills. But I refused. Maybe driving wasn't for me.

But the summer was coming to an end and school would be starting soon. Faced with another school year without a license and the embarrassment of telling my friends that I'd failed my driving test twice, I decided to give it another shot. I made an appointment for two days before my permit would expire.

I had barely practiced, so on the ride back to the DMV, with my mom again in the driver's seat, I was not in high spirits. We were in her Ford Taurus, as I didn't even feel the need to get the Metro again. Last time, it was a hopeful formality: let's just take this stupid test so I can get this stupid license. This time, it was a fatalistic formality: let's just take the test so I can fail it and be done with it.

By now, I had become familiar with the staff at the Colonial Avenue DMV, and the same DMV official who sat in on my first test got in the car for test #3. I didn't take this as a good sign, and my expectations sank even further; perhaps instead of driving up to the designated area to parallel park, I should just floor it and run over the cones, then get out of the car, slam the door shut, and scream, "There! Are you happy? That's what I think of your fucking cones and your stupid test!" Maybe even kick the car and stomp away.

But I collected myself. I drove up, stopped, and backed into the parking spot. Then I pulled out of the parking spot, left the DMV center, and drove around. I pulled back into the station and stopped. Then I was handed another slip and told, "Congratulations."

I'd now have to change my life motto. "Expect nothing, and you will be disappointment-proof. And if you succeed—hey, would you look at that."

For the next week, I was the belle of the ball. It was all accolades and congratulations. It had certainly looked hairy there for a while, but now, back at school, a newly minted junior, I was officially licensed to operate a motor vehicle by the State of Pennsylvania. Badass.

My mom was happy that I wouldn't be hitting her up for bus tokens at age forty-one. Kyle and Flem were pumped because now it was my turn to drive aimlessly around the suburbs of Philadelphia looking for house parties. But no one, including me, was as excited about my getting my license as my dad. My father and I finally had something in common aside from our gender and our last name: we were drivers. The biggest love in my dad's life was cars. And, I guess, his children. But also: cars. Up to that point, his eldest son had not appreciated this love, as he was too busy appreciating eating hot dogs and taking long naps. But now, maybe having a driver's license could be a turning point. Maybe his son was a car guy but had never realized it. Maybe now we two would spend hours in the garage, talking torque and transmission fluid or, I don't know, other cool-sounding car terms. Or maybe it was just that he was happy his son wasn't an embarrassment and finally had a goddamn license.

A few Saturdays after I got my license, my dad called and said he had a gift for me down at the pier and asked if he could pick me up during lunch. My dad was a longshoreman, but he was also a mechanic. In addition to the normal longshoreman duties (whatever those were), it was up to him and his crew to fix whatever was wrong with the machines and forklifts and trucks that kept the pier humming.

My dad had always been a fixer and tinkerer, something he picked up from my grandpop. My Grandpop Mugs was known for going for a walk to the store and coming back with, for example, a broken refrigerator, which he would fix up and either keep, give away, or sell. My dad was the same. If something was broken, he'd want to fix it, whether it was a car, a home appliance, or something he'd never seen before. In my case, when something's broken, I just buy another one—AND DON'T TOUCH THE BROKEN THING BECAUSE IT MIGHT EXPLODE SO JUST LEAVE IT THERE AND LEAVE IT BE AND DON'T EVEN GO NEAR IT.

I had a pretty good idea of what this gift was. I figured I was getting my first car. I assumed it would be a clunker, salvaged from some scrap heap, possibly two different colors but, knowing my dad, with a working radio and an engine that would get me where I needed to go. And I assumed it would be American, since, according to my dad, if you bought a foreign car you might as well go ahead and join the KGB and just get the whole thing over with.

After my dad picked me up and we arrived at the pier, we walked into the garage where he worked. I saw a long row of machines and forklifts and cars in the shop, all in various stages of repair. As we walked along, I wondered which one was mine, thinking that at any moment, my dad would stop, point to a car, and say, "She's all yours" in some real macho dad way, just like you see in car commercials. But instead we kept walking and walking all the way down the row, until we got to the office, a trailer at the end of the garage.

My dad said, "Over here," and we walked to the side of the office to an area no bigger than a shed, where my dad pulled off a blue tarp to reveal my present. It was a motorcycle.

It was a fucking motorcycle.

"Congratulations on getting your license."

It was a fucking motorcycle.

Up until I was in seventh grade, my dad had a motorcycle. And he loved it. And *I* loved it. When my brother, sister, mom, and I were living at my grandmother's house while my parents went through their divorce, my dad would sometimes pick me up at night, put me on the back of his motorcycle, and take me for a spin. We'd ride around the neighborhood for a little while, but our ride always had a grand finale. My dad would steer the bike to a desolate stretch of Delaware Avenue, which was then lined with abandoned buildings but now has a Home Depot *and* a Lowe's, and about four Burger Kings. We'd stop, and he'd rev the engine. And then we'd take off down the street, doing well above the 35-miles-per-hour speed limit, as I held on for dear life and shrieked with joy. It was like riding a roller coaster, only I was holding on to my dad, who happened to be *driving* the roller coaster. Though it only lasted a few seconds, it was the most exhilarating thing I had ever done. Hell, it's probably still the most exhilarating thing I have ever done.

I realize that if someone had seen our little drag races, someone like, for example, a Department of Family Services official or my mom's divorce lawyer, it would not have been a good

The possibilities were endless. Predictable, but endless.

thing, either for my dad or myself. But that was my dad. Many kids, when their parents are getting divorced, will get to eat extra ice cream and stay up extra late when they hang out with their dad. I got taken on drag races.

Then one day, the motorcycle was gone. When I asked him what happened to it, he said only that he "got rid of it." There were a number of vehicles that had come in and out of our lives, and I assumed this was because of the nature of my dad's job— that mechanics traded in cars every so often like other people might buy a new winter coat or a new pair of jeans. But I could tell that the motorcycle was different, and that when it was gone, he missed it. And for whatever reason, he never got another one.

Yet here, in front of me, was my very own motorcycle. And it was a real motorcycle, too, not one of the crotch rockets that you always see ninjas riding. Nor was it a glorified bicycle trying to pass itself off as a dirt bike. It was a blue Kawasaki KZ 650. Not a hog, but it had balls. It was used and it looked like the previous owner rode it from drag race to whorehouse to fistfight all across this great land we call America. If it could have, the bike would have looked over at me, snarled, spit on the ground, and said, "What are you looking at, pussy?"

I loved it.

My dad said he was still working on it but he couldn't keep it a secret any longer and wanted to show it to me. It would be ready the next weekend, and we could start riding it then. He then answered my unasked question, about the bike being non-American. "I picked it up for free from Billy Draper.* It needed a little work, but it'll be ready to go. And it'll have some balls."

I was stunned. It was the greatest gift I'd ever gotten. My friends universally agreed with this assessment. The consensus was that a) I had the coolest dad ever, and b) if I couldn't get laid driving a motorcycle, I would never, ever get laid. Despite my obesity, long hair, and clear braces, my friend Jim cautioned that upon getting the motorcycle, I should prepare myself for the hailstorm of pussy that would soon descend upon me. A hailstorm of pussy. Upon me. I agreed. One thousand percent.

But while the new motorcycle was lauded as the greatest gift in the history of gifts (teenagers category), I recognized that it was more than a present. For most of my life, my dad had been hands-off. And I mean this in a good way. He was not the type of dad to put me in a Harvard onesie as a newborn despite never having set foot on the campus. Nor was he the type of dad to

---

* My dad will often tell me the name of a person, a person I've never met or even heard of before, and expect me to know him. It's like he assumes that I know every person that he does, even if that person is some guy he shared the bus with in high school for a few weeks.

buy me a set of drums at age four because, man, he and his buddies had had a really tight rock band when he was younger, a band that he had to give up once my mom got pregnant with me so he could study accounting at the local community college to find a more stable gig. There was never any pressure from him to be or to like or to do anything a certain way. It wasn't like he addressed my birthday cards to "Justin": my dad worked a lot, but, aside from the occasional baseball game or outing here and there, he was around and available for consultation. But he was never all up in my bidness.

However, it was very clear to me that this motorcycle was, for the first time, an attempt to be hands-on. By giving me this motorcycle, was he trying to live vicariously through me? Or was it something else: his last-ditch attempt to man-up his son, to set me on the path of partying and pussy and steer me away from teaching pottery at the local women's center?

So as awesome as it was, it was a gift with a catch. This was more than a present for finally getting my license. This was a Hail Mary pass.

Though I had some bumps when learning how to ride a bike, I eventually mastered it. And though, more recently, I had some difficulty with my driving test, I was soon comfortably driving around on my own. A motorcycle was nothing but a cross between a car and a bike, I thought. So, really, how hard could it be?

This is what I focused on on our first day of motorcycle lessons. Once again, we were at the pier, in a large garage that looked like an empty airplane hangar. This offered some relief;

I was afraid that I might be taught to drive the motorcycle in front of my dad's coworkers. The pressure of my dad teaching me to ride was one thing, but lessons in front of a dozen dad-like (read: manly) mechanics was quite another.

I felt amazing when I first got on the bike. I had sat on my dad's motorcycle alone, and it had been fun to sit there and pretend to be driving. But this was not pretend. I was no longer a little kid making vroom noises and turning the handlebars this way and that. Soon I would be driving this bike. On my own. The idea was thrilling. It kinda made me hard.

Time is one of the few things that can change others' perception of a teenager once his reputation has been established. A quiet kid can grow up to be a rock god; a funny guy can turn into a serial killer; the star/stud quarterback can win a Tony for his portrayal of Raoul in *The Phantom of the Opera*. But teenagers often get pigeonholed, and that reputation usually sticks. Life is not the movies, after all; the "ugly" girl doesn't take off her glasses and immediately become the most beautiful girl in the school, nor does the captain of the chess team pick up a football for the first time and throw it seventy yards. In high school, for better or worse (usually worse), you are what the majority of people think you are.

But when my dad turned on the engine and started revving up the bike while I sat on it, alone, I knew that this motorcycle was a perception-changer. I could picture myself, the once-unassuming guy known for his self-deprecation, rolling up to parties on his motorcycle. As everyone turned to look, I'd sit there on the running bike and rev the engine a few times, and then three women would instantly orgasm. Then I'd rev it a few more times, even louder now, and my penis would grow seven inches and my balls would become roughly the size of softballs (but much, much harder), and then I'd rev the engine a few more times, loudest of all, and by now all sorts of hot chicks would be dropping in ecstasy all over the place. And then I'd nonchalantly shut the bike off, step off of it, and walk into the party like it was no big deal. Then I'd probably hit the bathroom to make sure I didn't have any bugs in my teeth. Because it had been kind of a long ride.

As the bike came alive, I realized that my balls were now on top of a several-hundred-pound machine capable of going over one hundred miles per hour. It was like sitting on a thunderbolt. How could this not be a game-changer?

My dad explained that riding a motorcycle, like riding a bicycle or driving a car, came down to three mechanical functions: go, stop, and steer. The last was the easiest, as the handlebars move the motorcycle just like on a bicycle. I quickly learned that the other stuff was not as easy.

I won't get into too much detail about how a motorcycle operates. I couldn't understand it then, and I don't understand it now. (Also, this is not a "How To Drive a Motorcycle" book.) But what I was looking (or hoping) for was a handle or pedal for the gas and a handle or pedal for the break: one to go, one to stop.

I learned to drive a car with an automatic transmission. I got in, turned the key in the ignition, and pushed the gas pedal to go. The harder I stepped on the gas, the faster the car moved forward. When I wanted to stop, I eased up on the accelerator and applied pressure to the brake pedal. The harder I pressed, the more quickly the car stopped. Pretty straightforward stuff.

This was not the case with the motorcycle, which had a manual transmission. Cars with a manual transmission were unheard of where I grew up; in a city, with a stop sign at the end of every block, it wouldn't make much sense to have to shift gears every five hundred feet. So when I got the motorcycle, I couldn't drive a stick-shift car.

In addition to the manual transmission, the acceleration and deceleration of a motorcycle requires the use of all of one's limbs. In a car, you use your right foot for both the gas and brake pedals. So that's one limb for both going and stopping. On a motorcycle, your left hand controls the clutch (I had no idea what a clutch was, and it needed to be explained to me). The right hand is the accelerator. The left foot shifts gears. The right foot is the brake for the back wheel. And oh, I forgot, you also have hand brakes. And, oh, you have to look out for traffic and to make sure no one throws an unopened can of soda at you while you're driving.

I could get the bike to move by letting up on the clutch. That much I could handle. But what came next—the left hand slowly releases the clutch, the right hand provides the throttle, the left foot gets ready to shift gears—was like learning how to dance, except that if you screw it up, you would careen headfirst into a tree or off an overpass. A few weeks earlier, I couldn't park a Geo Metro in a spot nearly double its size. Now I was expected to operate a motor vehicle that required the dexterity and coordination of a hockey goalie.

Despite my dad's efforts and best intentions, this was just not going to happen. I think it was better that way. What could have been a long, painful process of dad repeatedly trying to teach and son repeatedly failing was nipped in the bud. There was no hope that maybe it'd all suddenly click for me or that maybe, with enough practice, I'd figure it out. It only took a few lessons at the pier of me starting, stalling, nearly falling over, then starting all over again to realize that there was simply no chance that I'd learn to drive it. My father and I both knew that this was less of a "sorry that I disappointed you, Dad" situation and more of a "Dad, what in the world were you thinking?" situation. Yes, I could ride a bike and had just learned to drive a car. But asking me to translate that knowledge and experience into riding a motorcycle was like asking a freshman astronomy major to save the planet from a giant incoming asteroid.

There was only one thing to do: sell the motorcycle. I certainly couldn't keep it, and my dad didn't want it either. He said it was because it was a little small for his taste. I also assumed its presence would be a daily reminder that his first-born son was a total wuss. (He didn't say this, but I assumed it.)

My friends were bummed that it didn't work out, but not entirely surprised. Really, how could they be? Me driving a motorcycle from party to party and make-out session to make-out session was a fantasy. I put the word out around school that I was selling my motorcycle. The very next day, my buddy Mike told me his older brother, Kevin, was interested. Could he come take a look?

I liked Mike. He was first-generation American, the son of Irish parents, and one of seven kids who was growing up in a modest home in Northeast Philly. He was a funny guy and a bit of a lunatic. I said that sure, his brother Kevin, who had just turned eighteen, could come to see the bike whenever he wanted.

A few days later, Kevin and his buddy Tim stood at the pier with me and my dad, checking out the bike. Kevin looked like Mike, with his long, scraggy hair and easy smile. As my dad told him the bike's specs, Kevin walked around the motorcycle, looking it over studiously, grunting to himself, and occasionally touching parts of it.

When my dad was finished, Kevin looked at us and said, "I gotta be honest. I've never ridden a motorcycle before." He added, "But I really want to."

My dad beamed at the chance to show off his knowledge of the art of motorcycling once again. He gave Kevin a quick lesson, teaching him in about seven minutes everything he tried to teach me over many afternoons and evenings.

The lesson over, Kevin hopped on the bike and turned it on. He got a little run going, making slow, cautious turns around the garage. Then he sped up a little bit, making faster, less cautious turns around the garage. He wobbled a little bit here and there, but he stayed on. He brought the bike to a stop in front of us. "I'll take it," he said.

He handed my dad three hundred bucks and we all shook hands. Kevin's friend Tim got in his car and Kevin waved him along and then goodbye to us. Kevin revved up the bike and off he went, driving back to Northeast Philly. It was about a seventeen-mile trip and would require him to drive on the highway. And he was on a motorcycle for the first time. Without a helmet. And he didn't seem to be too concerned about either.

As Kevin pulled away on the motorcycle, I thought I saw a tear in my father's eye. Not so much for the bike that he'd worked to restore to working condition, but for the son that he'd never had, driving away and out of his life.

But the good news was that at least I could still legally drive a car.

the summer of
magical drinking

North Wildwood, New Jersey, was the most magical place on earth. An island beach town about ninety miles southeast of Philadelphia, it was where, for two weeks each summer, once in June and once in August, my family would spend our vacation. And when I say "my family," I include a number of people outside of my immediate family; my grandmom and mom's sisters, Anne and Maureen, and their families would usually join us. We'd get two adjoining rooms at a motel with a pool (and hopefully a waterslide) and cram ourselves in, kids sharing beds with parents, siblings sharing beds with each other, and, inevitably, my brother and I sleeping on the pullout couch. Dennis and I would get very little sleep on the couch, however; our Aunt Maureen was a serial sleepwalker, and her occasional ghostly nighttime walkabouts scared the shit out of us and kept us on edge all night long.

Despite the lack of a decent night's sleep, it was always a week of decadence, kid-style. We'd run the air conditioners twenty-four hours a day, something that was verboten at home, where my mom's desire to conserve electricity meant we were allowed to open doors just enough to slither in or out of a room, lest all the cold air immediately seep out. We'd spend entire days swimming in the pool, diving in when it opened first thing in the morning and getting yanked out by our parents when it closed in the evening. And we'd eat ice cream every single night. That wasn't much different from our life back in Philly, but down the shore we went out to ice cream parlors for our treats instead of eating them at home. *That* is how you vacationed in our family, baby.

Furthermore, everyone in our part of Philly went to North Wildwood, making it more or less Second Street-by-the-Shore. I don't know how everyone in our neighborhood came to spend their summer vacations in the same New Jersey shore town. After all, there were dozens of shore towns within easy driving distance of South Philly, a number of them even closer than North Wildwood. But going back generations, North Wildwood was the beach destination for Second Streeters.

This made going down the shore even more fun. It was not the type of vacation during which you were trapped with your family, the one week per year when your siblings had to be your friends because there were simply no other kids around. Half of my friends spent their entire summer down the shore, and the other half, like me, usually went down for a week or two each season. So going down the shore meant being reunited with those buddies who spent their summers in North Wildwood and whom I hadn't seen since school had ended.

My favorite part of our summer vacation was the boardwalk. The Wildwood boardwalk was as close to paradise that someone between the ages of three and twelve could possibly come. There were the multitude of games that bankrupted parents for the sake of winning a stuffed animal. There were the rides like the log flume or the Condor (which lifted you up four stories high, spun you around, and then dropped you down) and the Sea Serpent, the marquee roller coaster. There were the arcades with rows and rows of video games and skee-ball. And, of course, the food. Funnel cake and french fries and frozen yogurt and pizza . . . paradise, I tell ya.

Down the shore with (you guessed it) a few cousins (see if you can pick them all out). Also, this is the last known picture of me shirtless, as things got heavy once I hit age seven.

While it sounds like a Kids First! vacation, North Wildwood was just as charming a place for teenagers. The pool and the beach were viable and fun options regardless of age, but there was a new allure to this Jersey Shore town once kids got too old for boardwalk rides and games: new girls (and guys). While North Wildwood was Second Street-by-the-Shore, it was not filled exclusively with people from our South Philly neighborhood: there were families from the rest of New Jersey, other parts of Pennsylvania, and New York who summered there. Then there were the locals who made their homes in North Wildwood year-round, as well as the Europeans, mostly Irish and Polish, who worked in the town's service industries during the busy summer season. Add it all up and the opportunities for meeting someone for a little summer lovin' were almost limitless. Maybe your friends in the neighborhood knew you as the local miscreant, but the girl from Krakow working in the arcade had no idea you liked to start fires in neighbors' trash cans, and the guy from Larchmont who did Tuesday through Thursday nights at the ice cream shop was unaware that you'd peed your pants in gym class in fifth grade.

And if, over those high school summers, a teen from Second Street was unable to succeed in initiating a summer romance, there was a light at the end of the tunnel: the post-graduation shore house.

Every society has its rites for the transition from childhood to adulthood. Aborigine boys spend months alone living in the wilderness. In Turkmenistan, young boys and girls are expected to successfully wrestle and pin a wild pony before being considered an adult. In certain parts of Chile, sixteenth birthdays involve intense wine-drinking competitions between the birthday boy or girl and the local village wino that sometimes last for weeks.* For generations, one of our neighborhood's coming-of-age rituals was the Getting of the Shore House after High School Graduation. A half dozen to a dozen friends would chip in a few bucks each and rent one of the giant Victorians in North Wildwood for the whole summer. Everyone did this. Throughout North Wildwood there would be a rotating set of parties. The freshly minted graduates would spend the summer after their senior year working as bartenders or waitresses or lifeguards and drifting from one party to another, night after night, living the stories that would be told in bars and at parties twenty, thirty, and forty years later. The potential for summer lovin' was now off the charts; no longer were teens living with and under the terms and conditions of their parents. Many people who started dating during this special summer eventually got married (it was down the shore that my parents first "properly" met).†

The shore house upon graduation was a reward for the successful completion of high school. But it was also both a first time and a last moment. For most, this was the first time living away from their parents, and the freedom and opportunities it presented were exhilarating. Many of them would not be going to college and, if they did, they would not be living away at school and in a dorm. So this was a special summer to

---

* Please note that I made most of this stuff up. I googled "coming-of-age rites and rituals" and got, like, two hundred thousand results. I can't be expected to comb through all that. C'mon.

† Previous book plug #2: For an explanation of why I used the word *properly*, I refer you to *Everything Is Wrong With Me: A Memoir of an American Childhood Gone, Well, Wrong* by Jason Mulgrew. Available online and at fine bookstores everywhere.

enjoy the freedom of living with friends before moving back in with family and either getting a grown-up job or taking college classes. Eighteen years old, a dozen-plus years of school over with: it was time to party.

But I didn't have to wait that long. My friends Trusko, Big Rob, and I got a shore house after my sophomore year of high school. I was fifteen.

There are certain benefits to being good. That is, when you don't get in trouble and get decent grades year after year after year, you can use this unblemished record and stockpiled karma to your advantage one day.

That day came in the winter of my sophomore year of high school. By then, Steve Trusko had become one of my best friends. Known in the neighborhood as "Coast to Coast" Steve Trusko for his penchant for going coast to coast (read: never passing) on the basketball court, Steve was a rare bird like me. He had grown up on Second Street but also went to the Prep. Prior to high school, I didn't know Steve very well because he was a year older than me. But we bonded quickly on those morning commutes on the yellow Prep bus that made a loop around South Philly, hitting Second Street and a few of the Italian neighborhoods.

A bit of a troublemaker, Steve was a good guy who could win you over with his combination of charm and pure volume. If he was in the room, you couldn't help but notice it—you could hear him telling jokes and cracking up from about five hundred yards away.

It was Steve's idea to get the shore house. He knew that I

loved going to North Wildwood. He also knew that by now, sharing a pullout couch with my little brother and having a dozen family members jammed into two small motel rooms was getting a little old. And since I loved those two weeks each summer, I'd really love being there for an entire season, as Steve pitched it. Steve already had a place to stay down the shore, as his parents rented a place each year for the whole summer. But he, like me, was tired of sharing the place with his family.

He also knew that I worked and had a little bit of cash on hand. Steve, too, had a job during the school year and was able to contribute. Still, we would need a third roommate, as two dudes do not a party house make. Our third would preferably be someone whose company we enjoyed and who had money to spend on the rental. We looked to our friend Big Rob.

Like Steve, I didn't really know Big Rob until high school. He was the same age as Steve, and those two, like me and Floody, went back to when they were in first grade. Big Rob got his nickname because by the time he was ten or eleven, he was almost six feet tall and just under two hundred pounds. Now, a few weeks shy of his seventeenth birthday, he was almost six feet tall and just under two hundred pounds. Though the nickname didn't really fit—he wasn't big enough to be called "big," but he also wasn't so small that it was ironic—it stuck.

Big Rob would make an ideal roommate because in addition to being our good friend, he got almost anything he wanted. He had a car. He had the newest video games. He had the latest curfew. He was spoiled, but he'd admit it. Selling a shore house to his parents would be a piece of cake, something he agreed with when we revealed our grand plan.

And really, it wasn't hard for any of us to get our parents' permission to get the house. I don't think Big Rob asked so much as told, and Steve and I both pulled the "but I work so hard all year long, both in school and at work" routine. My mom was reluctant, but really, what was she going to do? I never got in trouble, I did well in school, I had a job and was paying for the shore house myself, and, I argued, that meant one less person crammed into the motel room come June and August. So, despite the fact that I was only a sophomore and not yet sixteen, my mom agreed and let me get the shore house. I barely had pubes, but for three months I'd live with two friends without any adult supervision.

We didn't need the standard grand shore house rental, with multiple bedrooms and bathrooms. All we were looking for was a small place, especially because two of us would be going back and forth from the shore to the city over the summer. I'd still work at Mick-Daniel's from Wednesday night until Friday night, and Big Rob, who had a car, said he'd drive back and forth to spend a few nights a week in Philly because he thought he'd be "bored" down the shore full time, which Steve and I took to mean that he'd miss his mom's cooking and her cleaning up after him. Steve would continue in his summer job as a dishwasher, which he'd had for a number of summers, at the local Elks club in North Wildwood, where his dad was a member.

We found the place through a friend. Our buddy Jerry's family owned a home, and in the back there were three connected low-slung bungalows. His great uncle lived in one, a second was undergoing renovation, and a third was an unrented two-bedroom, one-bathroom place.

Though we were young to have our own shore house, because Jerry's dad knew our families, he agreed to rent the place to us. This worked out for everyone. Our parents thought that because we weren't renting from a management company or faceless landlord, we wouldn't get ripped off. They also figured that the presence of Jerry's family on the property would curtail any excessive partying or bad behavior. It was a fit on both sides, so we rented the place for the whole summer. We officially had a shore house.

Our first weekend was the unofficial start of summer: Memorial Day weekend. I had to work part of the weekend at Mick-Daniel's, but I got a ride to North Wildwood, as I would every weekend that summer, from the bar's DJ, DJ Ray, after the bar closed on Friday night. I finished my shift by 10 p.m., went home, and slept for a little bit, and then DJ Ray called after 4 a.m., after he'd cleaned up, put all his gear away, and was ready to start the drive down the shore.

Though our relationship would develop over the coming months, I didn't know DJ Ray very well, and the rides were a little awkward. It was just the two of us in the middle of the night, the only car on the highway, both of us tired and not especially in the mood to conduct scintillating conversation. But a ride was a ride, and I wanted to get to the shore as quickly as I could after my shift ended on Friday night.

We got into North Wildwood just as the sun was coming up on that first Saturday morning. DJ Ray dropped me off at the curb, waved, and pulled away. I walked back past the main house to the bungalow area. From the street, I could make out a figure sitting on our small porch, smoking a cigarette. I found this odd, because neither Steve nor Big Rob smoked. As I got closer, I could hear music playing softly, and I recognized the figure.

It was, in fact, Steve, with a cigarette in one hand and a can of Schmidt's in the other. He was listening to Led Zeppelin's "Down by the Seaside," playing low on a boom box that sat in the window of the bungalow. Steve looked up at me, smiled broadly, and, with the cadence of someone who had just been hit with a tranquilizer dart, said, "Yo, cuz—welcome to our shore house."

He took another swig off the Schmidt's can and a drag of the cigarette, then stared off into space, making no further attempt to engage me—it seemed like he'd completely forgotten I was standing there. I walked into the bungalow and found Big Rob snoring on the living room floor, laid out like a homicide victim, with a can of Bud Ice within reach. The boom box and TV and every single light in the house was on, as was the whirring fan in the bathroom. The place looked like it had been ransacked.

I dropped my bag on the floor. Those first few minutes gave me a pretty good idea of what we'd spend most of the summer doing.

Neither Steve, Big Rob, nor I, nor any of our friends, were strangers to alcohol by the time we rented the shore house. By fifteen or sixteen, almost everyone was drinking. While our main hangout was the Park, we couldn't drink there, as it was a playground in plain sight of the surrounding row houses. So we started hanging around places that offered more privacy: under the bridge or on the dark side. The former was not an actual bridge, but rather the area under the I-95 overpass, which cut through the easternmost part of the neighborhood. While the west side of the overpass was residential, the east side was littered with old factories that were now empty and decrepit. Thus this was prime real estate for us, and there were no adults around to bother us or stop us from drinking, publicly urinating, and making out (not all at once).

The dark side was part of a schoolyard of a local high school. The school was fenced all around and took up most of a city block, but the school itself was shaped like a *U*. There was a schoolyard on each side, and a third one in the middle at the dip in the *U*. This was what we called the dark side, because (you guessed it) it was darker than the other two schoolyards. We could drink there in relative peace and quiet. We were not as undisturbed in the dark side as we were under the bridge—police cars were much more likely to ride through the dark side with their lights flashing to scare us away if we were too noisy—but the dark side was closer to civilization and less scary. (Though unsubstantiated, we had a fear of homicidal hobos attacking us under the bridge.) The dark side also offered easier access to bars and alcohol.

Though most of us were underage by a half dozen or so years, buying booze was as easy as buying soda. Not all the neighborhood bars were as upscale as Mick-Daniel's. There were numerous dive bars at which we could get served. I don't mean dive bars that serve the "cowboy special"—a PBR and a shot—for five dollars and have multiple Iggy Pop albums on the jukebox and a few Boston terriers hanging around. I'm talking about dive bars at which even if you were a regular you didn't go into the bathroom because you feared a sexual assault, regardless of your gender. Some people had fake IDs from their older siblings or cousins, but they were rarely needed. A few of the girls had, um, developed quickly, and it was those girls who would be our "runners," as in, "Jamie's gonna run for some beer. Who's in?" The girl would walk to the bar with some others and buy as many beers as she could carry, and as soon as she walked out, the rest of the crew would appear from behind parked cars to help carry the beers back to the dark side or under the bridge.

The biggest problem with buying alcohol was that old Second Street provinciality. You might not have a family member or family friend in one of the dive bars, but there might be a friend of a friend of a friend of your family's, or someone who recognized you as a Mulgrew or a Collins or a Brown. So we often had to walk out of the neighborhood to find bars at which we could get served. It was an inconvenience, but it meant little more than a longer walk back.

Our last resort for purchasing alcohol was asking kids older than us, but still not twenty-one, to do it. They were generally willing, as long as we bought them a quart or a forty for their troubles (quarts and forties being our quantity of choice).

Why did we drink? I can't say. Why does one climb a mountain? Because it's there and it's dangerous, and though there's a chance climbing it might kill you, you will almost certainly increase your chances of getting laid after you've done it. We were teenagers and we were bored, and we saw all our parents and our older friends do it; so we did it, too.

While in the city we drank outside, at the new shore house we could drink inside. We didn't have to worry about frostbite from holding a forty in a paper bag. Or about our moms catching us in the act while walking to Wawa. Or about having to pee outdoors on the side of a school or an overpass pillar. Or about the cops rolling up, blasting their sirens, and scaring the shit out of us. We now could sit down, talk, listen to music, watch TV, and pee in a toilet while we drank, just like the grownups.

This, as you might guess, made our shore house—and, for the first time, Steve, Big Rob, and me—immensely popular. Starting that Memorial Day weekend, our house was the destination spot in North Wildwood for our friends (and their friends). Many of the older kids who had just graduated high school had their shore houses, but these were exclusive and off-limits to anyone not enjoying their special graduation summer. But no one our age—high school sophomores and juniors— had their own house. We had cornered the market. It was like being the only 18-plus club in a town full of 21-plus bars.

Though we tried to control the crowds, we more or less had an open door policy, with people coming in and out all hours of the day and night. The place started getting busy during the post-beach hours, when friends would come by for a smoke or a beer before heading back to their parents' houses for dinner. Then there might be a lull. But by evening we filled up once again, our guests going through Marlboro Lights and various and sundry cheap canned domestic beer at a rapid pace. It was that first weekend that we hit our first snag of the summer: where and how to get alcohol.

It was more difficult to buy booze in North Wildwood than on Second Street. There was no chance of buying beer at the bars and carrying it out like we did in Philly. There were fewer bars, they were farther apart, and they were very strict on IDs.

To buy alcohol, you had to go to the only beer distributor on the whole island. It was just around the corner from our place, but they, too, were strict when it came to IDs.

Salvation came from an unlikely place: the older kids. The same kids who had tormented us or treated us with indifference all fall, winter, and spring now looked at Steve, Big Rob, and I in a different light. When they learned that the three of us had our own party house at which the "young heads" congregated to drink, they thought it was kind of cute. So they helped us out. Whenever they went to the beer distributor, only five hundred or so feet from our place, they'd usually pop in and ask if we wanted anything. And because these guys and girls were hitting up that beer distributor nearly every single day, as long as we had enough money on hand, we had a constant supply of alcohol.

Lots of fourteen- to seventeen-year-old visitors and an unending stream of alcohol meant we had to go out of our way to be respectful of our landlords. We had no choice but to be, since we lived in Jerry's family's backyard. There were two entrances to the house: one in the front just off the porch, and one in the back, through the kitchen. The back of our bungalow faced a house that was undergoing construction, so we used that as our main entrance/chain smokers' alley. We kept the front windows and door closed and the shades drawn, and we didn't allow any loitering out front. Still, people coming in and out all day long made us look a bit like a higher-class crack house. Did we succeed in covering up all of the underage drinking going on in our home? Of course not. It didn't take Kojak to know that a lot of young kids were getting drunk on the premises. But we figured that as long as we were decent about it, we just might be able to get away with it.

We immediately established certain house rules to maintain order, including but not limited to:

- You are welcome to crash if you are too drunk to go anywhere else, but no sleeping in either of the bedrooms, which are reserved for Steve, Big Rob, and me. We later amended this rule to: You are welcome to crash if you are too drunk to go anywhere else as long as you alert a parent of your whereabouts. This was after a night in early June when Laurie, one of my mom's good friends and a waitress I worked with at Mick-Daniel's, stormed into our place looking for her son, our friend Ryan, who had passed out on our couch after drinking a half a bottle of After Shock and eating a whole box of Cap'n Crunch.

- If you are presented with a once-in-a-lifetime make-out opportunity, you can use the smaller bedroom as a venue. "Once-in-a-lifetime" did not include making out with one's girlfriend or boyfriend, but did include one of our friends making a connection with a stunning member of the opposite sex at one of our parties. No one took advantage of this rule.

- To gain entry to the house, you must contribute something. Anything. Ideally, this would be in the form of alcohol, but it was understood that was difficult to pull off. Any food—ramen noodles, soda, cereal—was welcome. Failing that, we kept an empty

water jug by the door so that people could contribute pocket change or a dollar here and there.

- Any yelling after being told to keep one's voice
  down or blatant disrespect for the property and/or
  its occupants would result in immediate expulsion
  and possible banishment for the entire summer, no
  questions asked.

Our rules were simple but effective. Maybe we didn't enforce
them as much with some guests as with others—we weren't exactly ladies' men, but we knew not to demand pocket change
from a girl for shore house entry—but it all went smoothly. And
from that first weekend, we settled into our roles. Steve was the
bouncer, enforcer, and de facto leader. Big Rob was the sidekick/
strong, silent type. And me? I was the sober guy.

When I moved into the shore house, I hadn't drunk very much
at all. When "we" drank under the bridge or the dark side,
my friends usually did the consuming and I did the hanging
around. I'd had a shot here and there, and maybe a sip of beer
or a swig of a forty, but I had never gotten myself a drink and
consumed the whole thing. There were many reasons for my
sobriety. The generally accepted reason among my friends was
that I was a pussy. (I respectfully disagreed.)

The first reason that I did not drink was that, despite the
buzz about drinking beers and let's get drunk and oh my god
getting drunk is awesome, it didn't seem that cool. When you're
fifteen and you get drunk, 90 percent of the time you're going
to end up getting in trouble with your parents, getting in a fight

with a friend, or puking or peeing on yourself. (Actually, it's probably closer to 95 percent of the time.) Based on this, I was not in a rush to try drinking. Weighing the risks (listed above) versus the rewards (a drowsy feeling and social acceptance), it seemed like an easy pass.

This fed into the second reason why I didn't drink. I liked being a nonconformist. *Everybody* was drinking. Usually, to a teenager, the thing that everybody is doing is lame. At the start of the summer, I didn't realize that there were exceptions to this rule: that sometimes, the thing that everyone is doing really *is* cool, and that drinking was most definitely one of them. This would take me a few weeks to figure out.

My parents could not have cared less if I drank. My mom and dad agreed on maybe eleven things in the entire time they'd known each other (and I'm including things like "French dressing is the best salad dressing" and "Today is a nice day"), but they were both totally cool with me drinking, as long as I did so responsibly. Never drive. Be sure to call if you get too drunk. And that's it. Otherwise, I could have all the cheap beer I wanted.

Their attitude made me even *less* interested in drinking. A big part of my friends' motivation to drink was to break rules and to piss off their parents. If I drank, I wouldn't be breaking any rules or pissing off my parents. I would be doing what everyone else was doing. And I'd probably just get drunk and do something stupid.

And then there was the last and, for me, most important reason for my teetotalism: I didn't like the taste. All of the one-off shots I had taken over the years—gin, vodka, whiskey, rum—tasted so offensive that I didn't understand how anyone, anywhere, could enjoy them. Why not just drink rubbing alcohol, which was cheaper and could be bought anywhere? Beer was even worse. Everyone had already claimed a favorite beer by fifteen, calling themselves Bud guys or MGD fans or Coors Light* drinkers. But to me, they all tasted like bread soda.

By the end of June summer was in full swing, and our place developed more into a mission for wayward teens addicted to "ice"-named beers, wine coolers, and light cigarettes than a summertime rental. And it was awesome. People—girls!—were coming up to me at our parties, introducing themselves, commenting on how cool it was that we had our shore house, and how great the parties were. I felt like goddamn Puff Daddy. But inevitably, there would be an awkward moment in our conversation when a guest asked if I wanted a beer and I responded, "No, thanks. I don't drink." For the reactions I got when I said this, I might as well have said, "No, thanks. I'd like to murder you and skull fuck you in a closet instead."

By not drinking, I was preventing myself from taking full advantage of the incredible situation that had fallen into my lap. At best I looked like the uptight guy at the party, making sure no one breaks anything or puts their cigarettes out on the furniture, and at worst like the priest who runs the mission and tends to the flock. So I decided to throw myself into the scene and make myself a drinker.

---

* Coors Light cans were known as "tin hoagies." They got this name in Veterans Stadium, the former home of the Philadelphia Eagles and Philadelphia Phillies, where one could bring food but not alcohol into the stadium. So people would pack their coolers with cans of Coors Light wrapped in sandwich paper to pass off as hoagies to any security guards. I don't know why this term applied only to Coors Lights, but the nickname endures to this day.

I started experimenting with beer, as it was the most readily available. The first beer was gross. The second beer was gross. And the third beer was grossest of all. And worse, you had to keep going, as it took at least two or three more to really do the trick. No thanks.

Then I tried vodka on ice. This got me drunk quickly, but it was also cool: while the guests enjoyed beer out of cans, I drank my vodka on the rocks like it was no thang, even though the vodka came from a large plastic bottle that cost seven dollars. But the vodka rocks was even worse than the beer. Adding ice did not make it any better: it was like a long, slow shot.

Vodka and orange juice was a bit closer to my wheelhouse. But the prospect of drinking a half gallon of orange juice a night was neither tempting nor cost-effective, not to mention that I'd have diabetes by Labor Day weekend. So I remained without a go-to drink.

There was much joking among Steve, Big Rob, and me—mostly ball-busting directed at me—about always having a gallon of whole milk in the fridge, which was otherwise packed with beer and maybe some cheese and lunchmeat. I loved whole milk (never that watery skim stuff), and every morning for breakfast I'd have a big glass and a Tastykake. Then at night, I'd have a glass before bed to help me fall asleep.

It was Steve and Big Rob who creatively solved my not-drinking problem with a gift. Steve was now drinking Schmidt's, his dad's favorite beer, so frequently that a can of beer became a permanent extension of Steve's body. Big Rob stuck with Bud Ice because he "liked the can" and because it was our unofficial house beer, as it was strong and cheap. After one of the booze runs by the older kids, Steve plopped down a new bottle on the table to add to our collection, saying this one was all mine. It was Kahlúa.

He explained that the only thing that goes with whole milk is Kahlúa. And that it's delicious. He would know, because, he said, it was his grandmother's favorite drink, and she'd sneak him a sip when he was a little kid. "It tastes like a milkshake," he said.

The bottle was familiar to me, as I'd seen it in the bar. But I didn't recall anyone ever ordering anything with Kahlúa in it. But milkshakes? I liked milkshakes.

Steve made the first one for me: a little ice, some milk, and a bit of Kahlúa dribbled in. It did indeed taste like a milkshake. But it was booze: technically, I was drinking alcohol. I finished the first and had another, adding a little more Kahlúa this time. It made my magic milkshake taste even better.

The Kahlúa—and the Mug—were mine. I'm assuming I was holding the whole milk when this picture was taken.

From that point forward, Kahlúa and Cream (whole milk), or K & C, as I called it, became my drink of choice. It wasn't perfect. I couldn't really get drunk because my stomach started to hurt from consuming mass quantities of whole milk before the alcohol took effect, but I could get a nice buzz going. Though I had nothing to compare it to, the next-day feeling was generally terrible, as my head pounded *and* my stomach roared. And it was summer. We had no air conditioning, and a whole milk–based drink was not the most appropriate choice for the season.

Yet, inconveniences aside, K & C's were perfect for me. I was drinking alcohol. I was still being nonconformist, because no one else was drinking milk and booze. And finally, and perhaps most importantly, I liked how they tasted.

I spent the rest of that summer as a proprietor of one of the busiest party houses in North Wildwood. I entertained dozens of friends and guests. The older kids knew my name. Girls came into my home and sat on my couch. And I drank alcohol. It was a magical summer, just a few years early.

the first six loves
of my life*

Sandy Olsson
(Ingénue/Vixen, *Grease*)

Some young children are obsessed with trains or with Disney
characters. I was obsessed with the complicated love story of
Danny and Sandy and the wonderful world of Rydell High. As
a kid, I watched *Grease* on a near-constant loop. I knew all the
lines by heart, I could sing the songs word for word, and I had
(most of) the dance routines down cold.

If this obsession was, um, cause for concern for my Irish
Catholic "gay people, family problems, alcoholism, and anything that involves an open discussion of feelings do not exist"

---

* Technically, five. Don't worry, it'll all make sense in a minute.

family, they did not let on. In hindsight, I have my doubts that
my dad, uncles, and grandfathers, tolerant though they were,
were able to watch without prejudice the oldest male grandchild
in the family prance around, thrusting his hips, and pointing—
with his hair slicked back with water from the kitchen sink—
while performing "Greased Lightning."

Fourth-grade talent show. Only one loser actually dyed his hair black
and got the curl going. Probably the one obsessed with *Grease*.

But if there *were* concerns that my obsession with *Grease* was
the manifestation of the nascent stages of homosexuality, any
homo-fearing males in my circle had nothing to worry about.
For though I enjoyed the music and the story (and the camp),
there was one reason above all others that I watched *Grease* over
and over and over again: Sandy.

Before I knew anything about sex or kissing or the romantic
dynamic between boys and girls, I knew that Sandy made me

feel something that no one ever had before. It was the "Hopelessly Devoted to You" scene that really got me: Sandy, lovelorn and alone, stood outside in her nightgown, away from the slumber party, singing about the struggle between the heart and the head, conceding defeat, and resigning herself to her overwhelming feelings. I may have been four years old, but something somewhere in my belly said that I should go up to Sandy and give her a hug. She was too pretty to be sad.

Later in the movie, when Sandy got all tarted up, I felt betrayed. I was in love with the real Sandy, who, as far as I was concerned, was perfect. That she changed to please Danny made me sad and mad and jealous and disappointed me. (I was very emotional as a five-year-old.)

I still cannot watch *Grease* or the "Hopelessly Devoted" scene without feeling those same emotions. Sandy, if I could climb into the screen and sit with you on those steps, I would. You stay just as perfect as you are.

Kara
(Girlfriend #1, Real Life)

We lived at my grandmother's house from about second through fourth grades while my parents went through their divorce; my dad stayed at our house. The situation was not all that great. For one, I was convinced my grandmother's home, a three-story row house across the street from a city park, was haunted. All night long it creaked and groaned, and on some nights it sounded like the wind was talking to you. That the house was across from a park did not help. This was not the type of park where kids could play on the monkey bars and roll around in the grass. It was a park where kids, especially kids of the chubby variety like myself, got their bikes stolen and were forbidden from entering after nightfall. My brother and I were convinced that once night came, all sorts of crimes and murders (perpetrated mostly by monsters) would occur there. The four of us—Mom, my sister, Dennis, and I—shared the third floor of the house, which had a small deck attached to it. I spent the better part of two years convinced that someone or something was on that deck at night, biding his/its time and waiting for the opportunity to enter the house and eat me.

But there were advantages, too. I hung out with my two uncles, Joey and Billy, who still lived in the house and were like big kids. And we lived close to Fernon Street, a small street a block away from my grandmom's house on which two of my mom's sisters and two of her cousins lived. They had children, and those kids had friends, and so on that small street, there were about twenty kids around my age, and we all became friendly.

Even after the divorce, when my dad moved to an apartment and my family and I moved back into our old house less than a mile away, I kept coming back to hang out with my friends on Fernon Street. Among them was my buddy John, who was a year younger than me. I liked John because he was always up for some mischief. And I liked John because I thought that his sister, Kara, who was my age, was the prettiest non-Sandy girl I'd ever seen.

Kara had hazel eyes and dark hair that came to her shoulders, hair that was not really straight but not really curly. Kara was skinny and she was shy. I was fat and not shy. It was a match made in heaven.

Through a complicated courtship ritual that culminated in my asking her to slow dance at a local weekly kids' dance, Kara became my girlfriend. In sixth grade, the concept of "boyfriend/girlfriend" was nebulous, but the one nonnegotiable criterion was that at each one of those dances, which were held at a local Mummers club every Friday from 5 to 9 p.m., Kara and I would slow dance together. Otherwise, she and I spent the evening apart, talking to our friends of the same gender, coming together only when the DJ played "Love Takes Time"* or "Eternal Flame." The dances were the prime showcase for being a couple. Afterward we hung out outside of these dances with our larger group of friends, and my primary function as boyfriend was to walk her home and give her a hug goodbye. A quick hug at her doorstep, usually with John either next to us or looking at us through the window, was nothing compared to slow dancing for three minutes a couple of times in a few hours.

At first, Kara and I danced holding each other at arms length. This is how all the new couples did it. She'd knot her hands around the back of my neck and I'd rest my hands on her sides just above her waist, desperately wondering if she could feel my hands, sweaty from nerves, on her skin through her shirt. We were so far apart that you could have driven a motorcycle between us, but I was thrilled: this was me and my girlfriend, slow dancing here.

Then, at a dance a few weeks later, when the DJ played a slow song, Kara and I met each other on the dance floor and she put her hands around the back of my neck and gave me a little tug toward

---

* Mariah Carey's "Love Takes Time" was our song: it was the first song that we danced to together. Even today, when I hear this song, my hands immediately get sweaty.

her, stepping into me as she did. I slid my hands from her sides and around her back and knotted my fingers. She put her head on my chest and we hug-swayed to the song. That tug, her stepping into me, my putting my hands around her back, her head on my chest—I wanted the world to stop, to freeze, so we could stay like that for the rest of time and space and infinity.

On the walk home that night, Kara and I kissed. It was the first time I had ever kissed a girl. I had been thinking about kissing Kara for a long, long time. The more I thought about it, the more I became convinced that a) I really, really wanted to kiss Kara and b) I was really, really nervous about kissing Kara. But when we stopped at her door that night, she turned around and, without saying anything, kissed me. Not a peck—a real, open-mouth kiss, right there, right in the middle of the street, right in front of her parents' house (!!!). She stopped kissing me and pulled away, said goodnight, and walked into her house. I stood there for anywhere from one minute to eight years. I could have stayed there forever.

Jennifer
(Entertainer, *Mickey Mouse Club*)

In sixth and seventh grades, my friend Ernie and I were big fans of *The New Mickey Mouse Club*. If you're thinking to yourself, "Sixth and seventh grades? Isn't that a little old to be a big fan of the Mickey Mouse Club?" you would be correct. But Ernie and I were not so much interested in the singing and the dancing and the skits (or so we told ourselves) as we were in two of the female Mouseketeers. For Ernie, it was Brandi, the Cajun brunette. For me, it was the bubbly blonde, Jennifer. We don't need to analyze this one to death: Jennifer was my preteen Sandy. Now old enough to feel sexy feelings, I directed them toward Jennifer, who I thought was just the tops.

As with Sandy, there was even a defining moment in our relationship, an awakening when I realized that—holy crap!—this was the girl for me. It was a skit in which Jennifer, decked out in a formal gown, reenacted a lounge act, performing with a tuxedoed piano player, singing about all the crappy costumes she had to wear in sketches (compared with the less-crappy costumes of the other Mouseketeers). It was that skit, and her poise and her pipes, that turned me on to Jennifer. The song was called "Hate to Wear the Stupid Costume Blues," and if pressed, I bet I could remember 90 percent of the lyrics.

. . .

I've said too much. I feel like this is getting a little weird. Now that we've established that I have a thing for blond singers, let's just move on.

Celeste
(Entertainer, *10000 Anal Mani-acs 1, Breastman's Ultimate Orgy, Plan 69 from Outer Space*, etc.)

There was a stretch in my teens during which I was masturbating about as frequently as I was urinating. Celeste had a lot to do with that.

There were more popular porn stars. Celeste was a contemporary of Jenna Jameson, who was then (and still kind of is) everyone's favorite porn star. Another contemporary was Chasey Lain, who was immortalized in a song by those morons who did the "Do it like the animals on the Discovery Channel" song. Celeste did not have an angle or a hook. She was not known for being a specialist (e.g., anal or gang bang scenes), she was not Asian, she wasn't even blond. Yes, her fake breasts were so large that at times of mental clarity, free from lust, one might look at them and think, "My goodness, those must be painful to carry around all day," but that was par for the course for mid-90s porn stars.

But Celeste had . . . something. She was classy. I realize that this is a strange way to characterize someone who spent a significant portion of her workday cleaning semen out of her hair, but that's how I felt about her. Something about Celeste's scenes, like her breasts, felt unnatural. This isn't to say she was blasé while effing or a bad actress—she really made a go of it in each scene and made the viewer feel a plethora of emotions, all of them originating in the genital area. But she was clearly acting.

The same could not be said of other porn stars. Compared to Celeste, the other porn actresses were obviously just nymphomaniacs with severe daddy issues who figured out how to get paid to fuck. There was no art or thought behind their craft. Celeste seemed above the mindless fucking; she was more intelligent, more calculated than the rest, as if she had chosen porn not because she loved to have sex (which she did) but because it was a smart career path, one that would maximize her earnings and allow her to pursue her other, non-fucking-related interests. While other porn stars went out to clubs on the Sunset Strip and fucked Mötley Crüe and/or Mötley Crüe–type bands, one could imagine Celeste, after a hard day of filming, going home to her art collection or snuggling up in a chair to read Kant in his native German.

For these reasons, if there were a record of all the orgasms I've had in my life and who was responsible for those orgasms (kind of like how iTunes keeps track of how many times you've listened to a song), Celeste would be at the very top of that list. Perhaps someday my wife will overtake her, but she'll have a serious uphill battle.

Amber
(Girlfriend #2, Real Life)

Amber and I had been friends since we were kids. I had harbored a crush on her from sixth through eighth grades, but I eventually moved on because she began dating my good buddy Joe, and a man had to respect boundaries. We remained good friends, even after she and Joe broke up.

The summer of the shore house, Amber and I started dating. I don't know how this began. I am guessing there was a fair amount of the ol' Kahlúa and Cream (on my part) and a number of Miller Lites (on hers) involved. One night, we kissed. The next night, we kissed. The next night, she was my girlfriend. This whole shore house thing was amazing.

I am grateful to Amber for three things:

Starting me on a lifetime obsession with breasts.
There is no delicate way to put this: Amber had huge boobs. Gigantic. I do not say this to be crass or to brag. ("Dude, my high school girlfriend? She was *so-o-o* fucking hot. She, like, coulda worked at Hooters.") It is simply a matter of fact that Amber had some big 'uns. It ran in her family: her older cousins were similarly blessed.

Our relationship, even by Irish Catholic high school student standards, was incredibly tame. In fact, I did not ever see her boobs. Nor did I ever feel them, even on top of her shirt. If I did touch them, the contact was accidental, unintended, uninvited. That Amber was such a prude was perhaps a response to having the big boobs; the magnitude of her prudishness corresponded to the impressiveness of her bosom. That I had zero access to these boobs despite their spending *an entire summer pressed against me* was maddening. There they were, all the time, right before me.

Uncovering them became my singular goal that summer. Like a brilliant scientist who had contracted a strange, incurable illness, I threw myself into the boob cure. I would even joke about this with Amber, who, once we were close enough, proved she had a sense of humor about her boobs. But alas, when the summer ended, and with it our relationship (amicably, I should note), I had not touched any boobs (aside from my own).

But once our relationship was over, I could not simply turn off the breast switch. Just as a child's development can be altered by a traumatic experience, Amber and her bosom impacted me in deep, psychological ways, instilling in me a fandom for boobies that today remains unrivaled among my peers. Some men are leg men or butt men or have proclivities toward redheads or Asian women. I am a boob man. And it's because of Amber and that one (goddamn) chaste summer.

Introducing me to blue balls.

When we were not hanging out at our shore house, my friends and I would go out and sit on benches on the Wildwood board-walk and take in the nighttime traffic. Toward the end of the evening, couples would venture under the boardwalk, spacing themselves appropriately, to make out before moms or dads or other rides arrived to take them home.

After one make-out session with Amber, I noticed some discom-fort in my testes. This was new; I'd been hit in the nuts before and experienced that drop-to-your-knees-and-choke-on-your-tongue type of pain. But this was a deep, throbbing pain, right inside my balls. On the car ride home with Amber and her mom, I sat motion-less on the backseat, wincing with every bump in the road.

After they dropped me at my corner, I waved goodbye and waited until Amber and her mom were out of sight before slowly walking bow-legged back to the house. There, I found Steve, Big Rob, and an older kid, Grinch, playing cards and drinking beer in the kitchen. I immediately went into it: Something is seriously wrong, my balls are killing me, I can barely walk, I didn't hurt them or anything, it just started out of nowhere. Amber and I were making out under the boardwalk, but when we got into her mom's car the pain started. And they are really, really sore.

Though I was approaching a panic attack, Grinch was laugh-ing. When he stopped, he called me "moron" and "dickhead." And he diagnosed me: blue balls. It's when you get all worked up but don't get your nut off and so your balls hurt, said Dr. Grinch. And there was nothing to do but ride it out.

I retired to the bedroom to sleep it off, begging Steve, Big Rob, and Grinch to keep it between us. For the rest of the summer, the three of them called me "BB." I'm glad the nickname is no longer with me today.

Proving that a friend-to-girlfriend transition was possible.

This was the biggest lesson of all. Amber and I had been friends since we were eight or nine years old. For most of our friendship, I had a crush on her, a crush that did not seem reciprocal and that I did not pursue lest I compromise our friendship. And then, when we were sixteen, we started dating. This was a major breakthrough. It was, however, both a blessing and a curse.

As a kid, flirting with girls in grade school by making them laugh, I learned the hard way that just because a girl likes your jokes doesn't mean she wants to be your girlfriend. Many a valentine was politely declined; many dates were quietly begged off. So I stopped forcing the issue, burying any crushes or feelings I had for a female friend, contenting myself with our friendship.

But Amber was a girl who laughed at my jokes, who was my good friend, and who *did* want to be my girlfriend. It was confusing, but thrilling. If Amber could go from being a friend to a girlfriend, then perhaps such a transition was possible with other girls, if I could figure out the proper way to handle it and then things fell into place. Right?

And then there was . . .

my aim is true

Everything I needed to know about love I learned from an Elvis Costello CD and a teenage girl. No, this is not *that* kind of story. It does not involve a trip to Southeast Asia or a two-year bid for corruption of a minor. But we'll get to the girl in a minute.

The Elvis Costello CD was a greatest hits album with a green and black cover. I held it in my hands, contemplating where to put it in my new CD tower, which I was organizing so that the least-listened-to CDs were on the bottom (and so harder to reach) while the ones I listened to more frequently were on top. I had purchased this CD at the behest of Kyle, my otherwise hip-hop-worshipping buddy, who knew my tastes in music well and told me I'd love Elvis, an artist whom he also liked but whom his dad adored. I ordered the album as the eighth disc in one of those "eight CDs for a penny—and then six CDs for forty dollars each next year (but let's not focus on that right now)" deals. I had never listened to it, but I felt bad about putting it in the bottom of the tower, which would essentially banish it from my life forever, ranking it alongside the greatest hits of Steve Miller and the Eagles and their ilk. So I popped the CD into my Discman to give it a listen and see if I could more fairly rate its place in the tower.

After my Beatles epiphany, I became a kind of music slut, falling in and out of love with all the rock 'n' roll staples— Led Zeppelin, Jimi Hendrix, the Grateful Dead, the Who, the Rolling Stones, Cream and Eric Clapton, Queen, the Doors, Van Morrison, the solo work of Paul McCartney and John Lennon—consuming music at an alarming rate as I worked my way through Classic Rock 101. But though I got obsessed with each, my obsessions were fleeting and never came close to what I felt about the Beatles, my first love.

But saying that you love the Beatles is like saying you prefer being handed a wad of cash to being kicked in the balls. *Of course* you love the Beatles. *Everyone* loves the Beatles. It was impossible to be really intimate with the Beatles, because even if you learned everything about them, you'd still only be their 18,645,813th-most knowledgeable fan. Loving the Beatles was like having a crush on the most popular girl in school.

So as I made my way through the other classic rock bands and artists, I looked for my Next Beatles: something that would blow my mind and keep it blown, but that maybe wasn't loved in the same way by every non-deaf person on the planet. Page and Plant; Jimi, Jerry, and Bob; Roger and Pete; Mick and Keith; Eric: all had come and gone, leaving me momentarily enraptured but, ultimately, flat. I could still enjoy them and be inspired by their music, but they were no John, Paul, George, and Ringo. Not even close.

Which brings us back to Elvis Costello and the new CD tower. When introduced to Elvis that evening, I was vulnerable. I was wayward. And by the second verse of "Watching the Detectives," I was sure Elvis Costello was my next major crush.

He and I had major potential. Just look at him! He was a skinny, dark-haired, small, British version of me. (OK, so maybe we didn't look *that* much alike.) But he wore his nerdliness proudly. His songs were filled with clever rhymes and wordplay. (All of "Everyday I Write the Book" was so witty I could barely stand it, but "When your dreamboat / Turns out to be a footnote / I'm a man on a mission / In two or three editions"? C'mon.) He was musically gifted and able to craft a melody. (I sang almost nothing but the punchy, pop-y harmonies in "Oliver's Army" for weeks after first hearing the song.) Yet he was not just sugar-coated pop with smarty-pants rhymes. (Both "Pump It Up" and "Radio Radio" were "stand up and punch the air while you rock out" anthems.)

Finally, for all his versatility, Elvis could be a sad, sad dude. "Indoor Fireworks" made me want to be an adult, get in a serious relationship, and then break up and be devastated, just so I could properly appreciate the depth of that song. "I Want You" was the most intense piece of music I'd ever heard—vitriolic and desperate, there are public executions that are less traumatic and moving than that song.

I couldn't believe that one person could create such a range of music, each song so different from the next. He was a musical schizophrenic. I hadn't heard anything resembling this kind of depth and variety since . . . the Beatles.

Music had always made me feel things. When Led Zep-

pelin blues'd-out to "Since I've Been Loving You," I felt epi-
cally sad. When Jimi Hendrix ripped it up on "Voodoo Child
(Slight Return)," I felt I needed louder headphones. But the
songs of Elvis Costello made me both feel and *think* in a way
that no other music had before. His songs were not something
you listened to and left behind. They were works of art that
required thoughtful study. Each time I came back to them,
I heard or learned something new about them. And in the
process of learning more about them, I learned more about
myself.

I knew I still had room to grow, but I also knew that Elvis
and I would be happy together for a long, long time. The greatest hits CD had twenty-two songs. By the end of the hour, I felt
comfortable using the word *love* to describe eighteen of them.

My favorite song? That would be "Alison."

Alison was the –est. She was the superlative. Alison was the
prettiest, the most athletic, the coolest. Her older sister was
an absolute knockout—and her older brother would knock
you out if you suggested such a thing. Her family owned
the best bar in the neighborhood (Mick-Daniel's, the one at
which I worked) and were active and considered leaders in
the community. It was like Alison was genetically engineered
to be the archetypical high school "it" girl, built in a labora-
tory in Europe by top scientists and then unleashed upon our
little teenage Second Street social circle to break hearts, kick
ass, and take names.

I met Alison in junior high. She had a mess of blond hair—
huge curls that I'd spend most of high school thinking about

swimming in. Though we became friends immediately, it was in high school that we became the closest of friends.

I was drawn to Alison for the same reasons that everyone else was. She was a force of nature. Teen girls are not typically known to be bedrocks of self-confidence, but if Alison ever felt insecure, she never showed it. This is not to say she was cocky or a bitch or a cocky bitch; in fact, it was her insouciance that made her so appealing. It was impossible for her to be unaware of the effect she had on the guys in the neighborhood, but she never let on. She was calm, cool, and collected—like a goddamned assassin.* She was also funny, a ball-buster, and a good friend. It was for these reasons that she became my best (girl) friend. And I was her best (male) friend.

We were a strange pair. The blond beauty who kept up with all the latest fashions while being captain of the basketball team, and the chubby guy who was now occasionally wearing a fur cape around the halls after school because he thought it was cool.† It didn't matter. We spent a lot of time together, hanging

---

* I know, dear reader, that you may be thinking that I said similar things about, or had a similar reaction to, Shannon, another cute blonde for whom I developed feelings. But my feelings for Shannon were not the same. Shannon had been ephemeral, and there is something inherently romantic about that which is fleeting. A relationship becomes all the more dramatic when you know that it will end sometime soon, for reasons beyond your control. I'd known Shannon for a few weeks; Alison I'd known for years. Shannon was like a shooting star, burning brightly for a moment but then disappearing back into the heavens. Alison was the motherfucking sun: there, all the time—and hot as shit.

† You're probably looking for more here, but I don't have an answer. A buddy who worked in the theater department found a big faux fur, Viking-style cape in storage, figured I'd like it, and gave it to me, and I began wearing it around the halls after school. The good news is that if I have a teenage child,

out, going to the movies or to neighborhood diners, laughing, talking, and having fun.

Me, in my cape, at a mixer, at which girls were present. Spoiler alert: The next chapter is not about me losing my virginity.

It might surprise you to learn this, but I had a huge, massive, overwhelming, unstoppable crush on her.

I know, I know. The overweight guy who is best friends with the hottest girl in the neighborhood develops feelings for her? You don't say! Is water also wet, the sky also blue, the earth also round? Yes, it's all true. In my defense, at one point or another I fell for just about every female friend I had. You see, I wore many hats in high school. I was a dutiful, if not very masculine, son. I

---

there is almost nothing that he or she will be able to do to weird me out. I wore a goddamned fur cape, for chrissake.

was a mostly disinterested brother. I was a dedicated-but-only-when-I-really-needed-to-be student, a committed employee, and a loyal friend, assuming it didn't require too much work. And to many of my female friends, I was their gay best friend.

Of course, I didn't know I was their gay best friend at the time. The concept of the GBF was something that I did not learn about until my twenties, probably from an episode of *Sex and the City*. But when I did learn about it, it made perfect sense. It was me, as a teen, to a tee (minus the actual homosexuality).

I spent most of my life girl crazy, in love with love, in love with every girl I met. In kindergarten, I made meticulous Valentine's Day cards for each girl in the class. In elementary school, the cards were store-bought, but each came with a synthetic rose (girls whom I really liked would get two roses). I always had several crushes going on at once; some big, some medium, some small, and the objects of my crushes could shift among levels on a daily basis depending on my own whims and how each had treated me that day. It sounds like a lot to keep track of, but I enjoyed it.

With the majority of my crushes, I expressed no feelings in anything other than Valentine's Day–card form. I may have asked a girl out here and there, but my success rate was so low that I figured it was best to not even bother. Instead, my plan was simple. If I liked a girl, I would befriend her and make myself not only available, but ubiquitous.

Need someone to chat with during recess? I'm your man. Looking for someone to carry your books home? Why, I happen to be going that way. Feeling sad because the guy you like likes someone else? Here's my phone number—call anytime.

It was over the phone that I did my best work and became a much more efficient and effective GBF. Just as gambling or drug addictions start small, my telephone addiction started with one little call from my friend Maria, who phoned me one night after school during freshman year to ask for guy advice. We talked for a half hour. The next night, Maria called again and put me on a three-way call with her friend Dana. The three of us chatted for over an hour. A few short months later, I had my own phone line with a dedicated phone number and spent about three hours a night on the telephone.

What did my female friends and I talk about? I don't really know. There were main themes; me providing advice on relationships or giving thoughts from a guy's perspective was a big one. But that would imply that I was useful exclusively because of my genitals*—that I was only good as a male sounding board. This was not the case. Many hours were filled with conversation about the vagaries of teen life: friends, family, school—as well as love.

Spending time on the phone with these girls made me feel fucking terrific. I was the goddamn social director, the king of the phone lines. Any way you cut it, I had the phone numbers of dozens of girls. Girls would call my home asking to talk to me. Maybe none of them was my girlfriend. Whatever. That was semantics. I liked girls. I liked getting attention from girls. The details would get filled in later.

And to be clear, I still had lots of buddies and an active, dude-centric social life. I was not getting manicures or going

---

* Not in the way that I would have preferred.

on shopping trips with my girl friends. Nor did I attend girls' nights out or slumber parties—regrettably. (If there were a record for ejaculating without any direct contact with one's genitals, I might have made a run at that record during an all-girls-plus-me slumber party.) I could hang out around the Park or under the bridge and talk about sports and music and wanting to touch titties like the rest of my guy friends. But I then went home and spent a few hours on the phone with girls, recapping the night's events.

It wasn't entirely clear to me what my endgame was. Perhaps I thought that by being consistently available as a shoulder to cry on, a giver of advice, and as a great friend in general, one day one of these girls would suddenly realize that she should start dating me immediately. Then we'd have sex and it would be amazing. What I didn't realize was that by insinuating myself with my girl friends, I was stripping myself of my maleness—and, more or less, becoming one of the girls. It was a cruel irony. I thought that the closer I became to my girl friends, the closer I became to a real, live hand job from one of them. In reality, from my girl friends' perspectives, that I had a working penis and testicles was becoming less and less of an issue.

Hence, the gay best friend. I was such an afterthought from a sexual perspective that for all intents and purposes I was no longer heterosexual in the eyes of my female friends. They'd tell me everything from secret gossip to shopping-trip recaps, who they liked and who they hated, problems and positives. And half the time, I'd be on the other end of the line with a boner. I'm not saying that I would touch it or anything. But it was definitely there.

The feelings I had for Alison, though, were above that. That is not to say that those feelings were boner-proof—far from it. But we had a deeper connection.

The phone calls with Alison were the longest, lasting well into the morning hours, held in secret, after we were told more than once by our parents to hang up and go to bed already. The nights out with Alison were the most frequent—even if she happened to be dating someone, which was often the case, there were still the movies and the diners and the hanging out after school. The spending time with Alison was the most fun. The comfort and ease we had with each other was remarkable, even more so considering the depth of my feelings for her.

She held an enormous amount of sway over me. In one conversation, she mentioned the name of her favorite men's cologne. I purchased that cologne the next day and wore it every day thereafter. In another conversation, she told me that she thought boxer briefs were "so much more sexy" than boxers. Though I was a committed boxer-wearer at the time, I then wore boxer briefs every time we hung out—though I looked less "sexy" and more "pudding in a Ziploc bag."

Over time, my crush on her became unbearable. High school was dragging on, and before I knew it, I'd graduate. I hoped to go away to college, and she'd stay in Philly to finish her senior year. It felt like we were approaching the end. Something had to be done. But what? The indecision was killing me.

After my relationship with Amber, I knew that a friend-to-girlfriend transition was, in theory, possible. With Amber, it was quite easy, actually. But of all the girl friends I'd had, Amber was the only one who bridged the two words and became my girlfriend. Was such a thing possible with Alison?

I thought about it for weeks. Should I tell her how I felt? Or should I just keep quiet about it? I weighed the risks against the rewards. Either way, the stakes were high. If I confessed my feelings and she felt the same, *BLAMMO*. I'm not saying we would get married right away, but I'd start calling around to check the availability of certain wedding reception venues, just to be safe. On the other hand, if I told her how I felt and she didn't feel the same, it would be catastrophic. At a minimum, it would compromise our friendship. Realistically, we probably wouldn't be friends any longer. Worst-case scenario: the whole neighborhood would learn what a fool I was and I'd have to start a new life somewhere west of the Mississippi. Maybe in Montana. I'd finally be able to put my knowledge of that state to practical use.

With the benefits (eternal wedded bliss) and the dangers (expulsion from the city) of a confession being just about equal, I had to make a decision based on another factor. Which outcome did I think would be more likely?

Alison could have her pick of any guy she wanted. That she was the prettiest girl in the neighborhood was up for debate as much as whether Christmas was a better holiday than Arbor Day. Her boyfriends were usually the best athletes and the best-looking. In my corner, I'd had one post-puberty girlfriend—for about two months. I was on year four of my clear braces. And again, the cape.

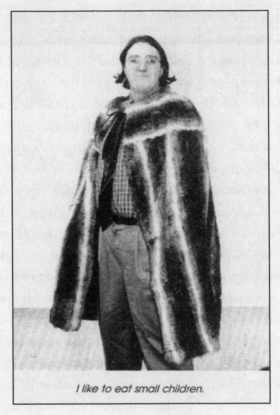

I like to eat small children.

Again, the cape.

It didn't look good.

I made my determination based on the idea that Alison must have known how I felt about her. We were great friends, yes. But cute girls didn't have non-cute guy friends in that way. Maybe she didn't want to acknowledge it, but her friends and family must have pointed it out to her. "This guy likes you—how do you not see it?" I could see her older sister telling her. So she *had* to know about my crush on her. And that she knew I was crazy about her but did not so much as even hint about it—and believe me, I analyzed every conversation for any shred of evidence of such a thing—was all the answer that I needed.

I decided to say nothing, to keep my feelings to myself. It was better to be her ever-loyal, functionally gay best friend than to lose it all. I knew that keeping my feelings to myself would be difficult, but the possible negative outcome of the alternative would be worse. And I'd leave for college pretty soon, so I only had to hold on for a little while longer.

A few years ago, I went on a blind date with a woman who said the strongest, purest feelings she'd ever felt were for someone in junior high. As I had already had a few beers, I understood this to mean that she was in love with a seventh-grader, and I told her that though I was open-minded about a lot of things, this was not one of them—unless he had some sort of developmental disability and was actually nineteen or twenty years old but had the cognitive abilities of a twelve-year-old. If that were the case, I'd be cool with it.

No, she said. The strongest and purest feelings she'd ever had for someone was when *both* of them had been in junior high.

"Oh," I said. "Oh. Right."

Alas, we did not see each other again. But what she said stuck with me. At first, I thought it was sad. How could an attractive woman in her mid-twenties not develop any greater feelings in her life than the ones she'd felt for someone in junior high? What an incredible head case she must be. And since she didn't call me back, what an incredibly good judge of character she must be.

But after spending some time thinking about it, I found myself understanding her sentiment. When we are young and have feelings for someone, we have those feelings without condition or consequence. There are no ulterior motives. It's not a case of "I like you because I find you really sexy and I'm going to a destination wedding in the Bahamas in six weeks and I'd love to bring you so we can fuck in the hot tub." Nor is it, "I like you because I'm getting pressure from my family and friends to settle down and start a family and you seem like you have few catastrophic defects in your family history and you have a good job, so you'll do."

And because we have no concept of what a real adult relationship involves, our imaginations are free to run wild. We don't realize that sometimes the people we date will lock us out of the apartment because we're on our way home drunk (again), or flush our cell phone down the toilet because they suspect us of cheating. In our minds, everything involved in dating is rainbows and roses and slow dances. When we're young and have feelings for someone, we have them because we just want to be around that person. That's it and that's all.

And when I thought about what my blind date said to me that night, I thought about Alison. And I thought about "Alison." And I thought about Elvis Costello singing the line "My aim is true" over and over again at the end of the song. For all my Elvis Costello scholarship, it only took me a dozen years to figure out what he meant.

236 pounds of
class vice president

As my junior year wound down, it was time to take stock of my high school career and tackle something I had been forcing myself not to think about until I absolutely had to: the eight murders I committed in 1989.

I'm just kidding, of course.* I'm talking about the college application process. After all, the point of going to the Prep was to get into a good college. It was the next logical step in my pursuit of world domination and my mom being able to tell everyone she knew, "My son Jason is going to [name of prestigious university]. Don't know if I told you that already, but he is. Yeah, so . . . what are you up to again?"

I didn't really have a preference as to which college to attend. I was getting advice from all quarters: Go to the best school you get

---

* It was only three murders. And I didn't so much "commit" them as watch them from the car and masturbate.

into. It's not the school, but the education. Go south or west to get out of your comfort zone. Stay in the Northeast because long-distance travel is such a hassle, especially over the holidays. Small schools are better. Big schools are better. Stick with the Jesuits. Get away from the Jesuits. The better the school is, the worse the girls look.

But my list of colleges didn't matter at that moment. I would worry about which school to attend when it came time to apply and again when the acceptance and rejection letters started coming in. Right now, I could only control my viability as a college applicant.

My grades were pretty good, but not great. Over the course of my three years at the Prep, I discovered that I loved and excelled in Classics, so I took a number of Latin and Greek classes. I don't want to say it was love at first sight—those first weeks of mastering *villa est villa Romana* were slow going—but there was something about the order and the structure of Latin and Greek that appealed to me. It was like cracking a code, and when you figured it out, everything was neat and orderly. I also liked that we Classics nerds were a small group. Out of two hundred kids in my class, there were only a dozen of us in Greek II. It wasn't snobbery. That we took these classes didn't mean we were smarter than other students: my knowledge of math never advanced beyond that last *144* square on the multiplication table. But in keeping with my code-cracking metaphor, we were the select few chosen to undertake this important task, like a group of specially trained soldiers (specially trained soldiers who had 1.5 sexual partners between them).

In most of my other classes, my grades were fine. Science was another matter. I had always been interested in big-picture science, stuff like explosions or transplants or experiments that turn a man into a hawk. But I was bored to shit by the micro-level stuff and showed little aptitude for it. Whatever hope I had for becoming a mad scientist was crushed during junior year, in chemistry.

There were two priests who taught chemistry. One was a funny guy and a relatively easy grader. The other was about one hundred fifty years old and, we guessed, had been friends with St. Ignatius of Loyola, Jesuit #1 himself. Though the classroom was small, he spoke through a microphone on his lapel, which projected his voice out of an amplifier on his lectern. He had the gravelly voice of someone who had smoked two packs a day for sixty years and the posture of one who still got in fistfights while in line at CVS. If Robert Loggia had a tougher, less ethnic uncle, it would be this priest. Worse, his grading premise was more or less, "You start with an F and must work your way up from there," as though he wanted to personally destroy not only your GPA and chances for admission to a good college but your life and your home as well. After a few weeks in that class, I swore off science forever.

The yin to the GPA's yang was the all-important SAT score, which took three years of intense schooling and a life-time's worth of knowledge and put them into one, easy-to-read number based on a three-hour test given on a Saturday after-noon. We were yentas when it came to SAT scores, sharing ours with a friend or two with the promise not to tell others (only to find out our score had made its way around the school by the very next period) and dishing about who aced or who bombed the test.

It became apparent that SAT-takers fit into three camps. The largest was made up of those kids who scored about what they should have scored on the SATs; their grades matched their scores. The two other groups were the outliers, either good or bad. Some kids with perfect GPAs scored well below what they should have, thus confirming our suspicions that they spent every waking moment studying because their parents beat them with a cane. Then there were the kids who got terrible grades who killed the SATs, thus confirming our suspicions that they had serious drug or alcohol problems. I was in the majority group, having scored about what I expected to score. I was fine with my score and only took the test once. The two most stressful experiences of my high school career (that did not involve contemplating professing my undying love to a girl who up to that point had assumed I was a homosexual) were taking my driving test and taking the SATs. Once I got what I came for from both, I never wanted to think about either again.

In terms of both my grades and my SAT scores, things were what they were by this point in my academic career. My GPA wasn't going to move drastically in either direction based on what I did my senior year. And even if I did take the test again and prepared hardcore for it, my SAT scores weren't going to increase by triple digits. As I endeavored to make myself more appealing to colleges, there was only one area in which I could improve quickly: extracurricular activities.

Extracurriculars were the weakest part of my game. I played no sports; failing gym my sophomore year had not inspired me to go out for track or pick up a lacrosse stick. (I did, however, effect a remarkable increase in my ping-pong prowess.)

I wasn't in many clubs. I joined Students Against Drunk Driving, but so did 95 percent of the student body. I think all you had to do was say "I am against drunk driving" and boom, you were a member. Because I hung out with some of the guys on the school newspaper, they bestowed the honorary title of Headlines and Captions Editor upon me. This meant that right before the paper went to press I was called in to read it over to see if I could come up with any funny captions. Mostly, I was not.*

I could always say on applications that I worked a real job for real money and that was my main extracurricular activity (and by doing so, queuing the violins and playing the "indigent kid" card—my understanding was that colleges loved that "overcoming adversity" shit). But it wouldn't hurt to buff up the resume with something high-impact, something that showed leadership and commitment to a cause greater than learning new ways to masturbate, but one that did not require years of toiling away in the yearbook office in order to finally rise to the position of editor or traveling from debate to debate semester after semester to become team captain.

I could do a service trip.

I could sign up to spend a week in some impoverished place to build houses and help locals and live in a hut and learn about a new culture. There were a number of options available through the school. If a service trip didn't look good on my

---

* My favorite was when Arlen Specter, then-senator from Pennsylvania and the former Warren Commission staffer credited with devising the single-bullet theory of JFK's assassination, spoke at our school. I captioned his photo, "You see, it was a special, *magic* bullet." This was my first and last foray into political satire.

transcript and give me something to write an application essay about, nothing would. One week away from home and I'd have my high-impact extracurricular over and done with. Perhaps I might even fall in love with a local native girl, who, after showing me flirtatiously the right way to prepare rice or fish or trees or whatever native people eat, would take me behind a hut and blow me as if the apocalypse were nigh. Yes, that might work.

But there was a problem with this plan. It would require me to go to some impoverished place to build houses and help locals and live in a hut and learn about a new culture. The learning a new culture thing I could get behind, but all the other stuff? Bleh. It sounded great, and I had little doubt it would look terrific on my resume, but, I mean, have you ever actually swung a hammer? Like, a bunch of times in a row? It's terrible. And let's be honest. My imaginary native girlfriend would prefer whatever jock was on the trip once she saw me faint upon encountering a bug or heard me shrieking in terror during a stormy night.

Fortunately, there was another avenue I could pursue, an extracurricular that would look terrific on my record and require very little work. And what little work it required would be fun: student government.

To be president, vice president, treasurer, or secretary of the whole school was a leadership position that would demonstrate my commitment to a greater, non-self-loving good. I had some experience in student government, as I had been a homeroom representative sophomore year, and I knew that to be an officer was not such a difficult gig. There was a monthly meeting, a few official appearances at freshman orientation and pep rallies—and that was pretty much it.

And there were some significant perks. The primary one was a key to the student council office, a small, windowless room tucked away on the second floor. It had a couch, a chair, and a TV with Sega Genesis. A man cave it was not, but it was the most exclusive clubroom on school grounds. The other perk was that as an officer of the school, you were able—nay, you were *required*—to go to the Prep's sister schools and sell tickets to our mixer dances during their lunch periods. Spending a portion of an afternoon at a table in the senior's cafeteria of an all-girls Catholic high school, seventeen- and eighteen-year-old girls in white button-down blouses and pleated skirts everywhere, with their little knee-high socks and their afieanofnwv;svnis;rv snvrhishw.[*]

The Prep ran its student government elections like most other schools. Candidates campaigned, covering the school with posters and buttons. The campaigning led not to a debate, but to a morning assembly at which every candidate addressed the junior, sophomore, and freshman classes for one minute. Then those three classes spent the day voting. (Seniors were left out of the equation because by that late in the spring, having already committed to colleges, they spent more time in the parking lot smoking cigarettes than in the building caring about student government.)

This was a much better resume-building plan. Any bozo with a week to spare, an iron stomach, and a genuine commitment to social justice could go to some godforsaken place in Latin America or West Virginia and directly and positively

---

[*] Sorry. I came while writing that—and also each subsequent time I tried to finish that thought. So you're just going to have to accept it as is.

affect the lives of dozens of people. But it took a real badass to run a successful high school student government campaign.

The first step was to pick which position I wanted to run for. Class president was immediately disqualified; it was the only position that required any real time commitment. Besides, the president had to give the commencement speech at graduation. No thanks. In theory, class treasurer sounded cool—being the guy who controls the money—but the position was really that of a glorified accountant. As for class secretary . . . who wants to be a secretary? That left vice president.

Class vice president was the highest-ranking position after president, and it required nowhere near as much work. Of course, there was a chance that the vice president might rise to the presidency if the president were caught in a scandal or otherwise had to resign, but if that happened, that would be cool enough to warrant the extra work (school newspaper headline: "Mulgrew Rises to Power, Jones Resigns in Disgrace amid Cheating/Cocaine/Human Trafficking Ring Allegations").

Also, our student council had two vice presidents, as opposed to only one officer in every other position. This changed the dynamic of the election. In the other races, friends often ran against friends and there could be only one winner, making for some awkwardness between buddies. But in the vice presidential election there was no such unpleasantness because any two candidates could win. In the other races, it was, "Vote for me!" which implied "over everyone else." In the vice presidential race, it was, "Vote for me—and any other qualified candidate! I don't care which one! As long as I get one of your votes, we're cool!"

Just like most high schools in America, the student council

elections at the Prep were more or less a popularity contest. About a third of the candidates had no shot at all. These were the super-nerds. Some were idealists who wanted to effect change but were clueless about their social standing at the school, while others signed up just so they could tell their overbearing parents that they had run. Another third of the candidates were fuckups, stoners, and/or partiers, some of whom were quite popular but all of whom very likely signed up under the influence of alcohol or drugs and had no desire to do any actual campaigning because that would mean less time for listening to Grateful Dead tapes and eating Taco Bell. It was not impossible that a member of one of these groups could win an election. Indeed, the treasurer of the current student council was a card-carrying stoner who was elected based on his promise to bring a Slurpee machine to the cafeteria (a promise, it should be noted, that he delivered on, because his dad bought and donated the machine to the school). But most of the victors would come from the final third of candidates: those with a legitimate shot at winning.

I thought I had a shot. I considered myself popular with the kids in my class and felt confident that, of the fifteen contestants for class vice president, I could win one of the two votes that my classmates had. But because of my lack of other activities, the only underclassmen I knew were the small group of kids with whom I took the bus. To reach the freshmen and sophomores, I had to take advantage of campaign promotional tools and turn myself into a motherfucking brand.

First, I needed a hook, a slogan, something catchy, something above and beyond "Vote Mulgrew for a Change" or "Mulgrew Will Deliver." A few years prior, a black candidate used the slogan

"Don't Be a Clown, Vote for the Brown" and swept into office in the biggest landslide in the school's history. My freshman year, a candidate with the last name *Bayder* somehow got the OK to use the slogan, "Vote for the Master: Bayder for Secretary" and had also won. I needed something similarly catchy or memorable.

One evening after school, I was lying in my bed, going over my options. There were a number of rhyming possibilities ("Mulgrew Is for You" and "Mulgrew Will Rock You"), but all were too corny and most vaguely sexual. So I started to think bigger picture. What did I have to offer the student body that other candidates did not? My opponents were star football players or wrestlers, involved in the campus ministry and community services, leaders in the Model UN and the Spanish Club. I was none of these things.

I stared up at my poster of Jimi Hendrix, kneeling over his flaming guitar, thinking. I really did have nothing to offer in terms of extracurriculars or my involvement in the school. After all, that was the whole point of running: to get myself a good extracurricular. I had no sports practices or responsibilities or obligations of any kind. If I were voted class vice president, it would be my only activity during my entire high school career. And maybe that was my hook.

All the other candidates had a lot of things going on. I did not. No practices, no meetings, no nothing. Once I woke up and made it to school, my day was pretty much obligation-free. So I could promise to dedicate myself in toto to serving as the class vice president. "Every ounce of my being would be your vice president." That might work.

I got off the bed and walked into the bathroom to see just

how many ounces we were talking about. I stepped on the scale, and there was my answer staring back at me. I, Jason Mulgrew, could promise you, my fellow students, 236 pounds of class vice president.

It was a hell of a slogan. But 236 pounds? I thought I had been in the neighborhood of "an even two hundred pounds of vice president."

During junior year, because of my strange class schedule, I had only one free period—seventh. There were eight periods in the day, and most people had lunch during fourth, fifth, or sixth period. By the time I got to my second-to-last period-of-the-day lunch, I was ravenous and gorged myself like a returning prisoner of war. It is no small miracle that I survived my junior year without losing a finger while "eating," which involved shoving food down my throat as rapidly as possible while grunting and snorting and sometimes barking. Because I was so hungry, I usually doubled up, eating two cheesesteaks or two chicken-finger sandwiches, washing them down with a soda or a Snapple and, if I had room left over, a TastyKake.

A few weeks into junior year, I could tell my clothes were getting a bit snug. But it was fine, I told myself. I had always been functionally fat, which I define as someone who still requires only one seat on an airplane, is able to kneel and rise or buckle his/her seatbelt without assistance, and is not winded by tasks such as clapping or stretching. And I had always embraced being fat. It had become a part of my personality (self-deprecating humor being the primary weapon in my arsenal) with which I made friends and influenced people. I was the funny fat kid, ever ready to call myself out before anyone else did. Everyone in high school has a thing. That was mine.

Yet knocking on the door of two-fifty was a wake-up call, too much of a good thing (or maybe just too much of a thing). But I'd deal with that later. Right now, I had a campaign to run.

For all its brilliance, it was not my slogan that gave the campaign legs, but my buddy Dan's "green" initiative. Looking at the campaign signs and posters that began to cover the walls of the school, Dan, who was not running for office, made an astute observation. "You know, it's all white. Every sign is on white paper or white posterboard. You should really use a different color, something that stands out."

His suggestion was so simple and so obvious that I immediately dismissed it as one of the dumbest things I'd ever heard. (I was also maybe a little upset because, though I planned to roll out my signs over the week leading up to the speeches, about 25 percent of them were completed—and on white posterboard, just like everyone else's.) The reason that all the signs were on white was because that was a campaign rule, I told him, having no idea whether or not this was true. I assumed it was; all posters and signs had to be approved by Mr. DiNapoli, science teacher and student government moderator, before they could be hung. This was to prevent students from hanging campaign posters with a picture of a pair of titties that said, "A Vote for Rogers Is a Vote for Titties!"

After school, as I was making the rounds hanging a few of the signs I'd already created, I stopped by Mr. DiNapoli's office. I brought up Dan's suggestion dismissively, saying what a silly idea it was, that there was no way it would be allowed, I mean,

God, what a crazy person Dan was. DiNapoli, finally looking up from the stack of tests he was grading, said that no, there was no law on the books that regulated the color of the posters—just their content. Campaign posters could be any color a candidate wanted.

I went to the library and found Dan. Swallowing my pride, I told him about my chat with DiNapoli and said that maybe he was on to something. And maybe we should go to Kinko's right away to put his strategy into practice.

An hour later, we were sitting at my kitchen table with every piece of neon green posterboard and every ream of neon green paper that Kinko's had in stock, as we'd decided that this was the color most likely to stand out—aside from hot pink, which would have sent a different kind of message at my all-boys Catholic high school. Dan and I spent that evening coloring and scissoring like a pair of kindergartners, transferring the completed white signs onto our new official campaign color. Because the white signs had already been approved, we could start hanging the neon green ones the next morning.

Dan's idea was remarkably successful. A week later, on the morning of the speeches, Jason Mulgrew was known as "the neon green candidate" more than "the 236-pound candidate." Other candidates may have had greater resources to spend on campaign promotion, seemingly employing whole teams of Central Americans to mass produce campaign posters, but everywhere you turned, there, among the sea of primary colors and pastels and pictures cut from magazines and stenciled names, all on white paper, was my bright-ass neon green with the slogan "Jason Mulgrew: 236 pounds of VP, every last ounce ready to serve" and a pithy tag, like "And that's a lot of ounces," or "Believe me, he's free," or "It's like getting two VPs for the vote of one." I promised Dan that, should I win the election, he would be duly rewarded with near-unlimited access to the student council lounge ("near-unlimited" because if I was making out with a chick in there during a mixer, which was sure to happen, he would not be allowed in).

I'm going to go out on a limb and guess that the caption for this one was something like, "Don't let your school go down the toilet."

But while promotion was important, it all came down to the speech. For many, the student government speech was the defining moment of their high school career. It was not often that a student got sixty seconds all to himself to address three-quarters of the entire student body. There was some potential for disaster. What was stopping someone from running an entire campaign just for the chance to stand before his classmates and expose himself?

Like everything else campaign-related, speeches had to be approved. The guidelines were reasonable. Candidates could have (approved) intro music after they were announced and (approved) outro music once they finished speaking, and they were allowed to use (approved) props. But once a speech was reviewed and given the OK, it was made clear that any deviation from the speech would not be tolerated and would be punished harshly, not just with automatic expulsion from the race, but with possible expulsion from the school.

That was not a worry of mine. I had no desire to show my genitals to seven hundred classmates and faculty members (not in such a well-lit environment, anyway). I was in a good position: my posters were visible and considered funny, and my campaign buttons—cheap name tags that were not much more than an index card with a pin—were in high demand and worn not just by the most influential of my classmates but also by the coolest members of the sophomore and freshman classes. A top-two finish was a real possibility. I had no desire to screw that up, because there was no way I was going on a service trip senior year to pad my unspectacular resume.

The speech was my opportunity to drive home my message: I can be a better vice president for you than any other candidate because I have nothing else going on. I also wanted to maintain my association with neon green. My original plan was to wear campaign pins covering my entire body during my speech, turning myself into a veritable green man. But I had a nervous stomach, and often had to use the restroom before stressful events. Pooping would be very difficult if every inch of my clothing had a pin on it, and I'd wind up leaving about two dozen pins on the bathroom floor should I have to go just before my speech.

So instead, I decided to cover the cape with the pins. I was already planning to wear the cape as I walked to the podium, then remove it dramatically to reveal myself covered in neon green pins, all while Jimi Hendrix's version of "Wild Thing" (it was a fur cape, after all) played through the loudspeakers. If the cape was covered in pins, I could still go to the bathroom if necessary and then make my grand entrance, only now I would shimmy up to the stage, James Brown–style, before letting the cape drop to the floor.

This was my plan as I sat in the front row with all the other candidates and Mr. DiNapoli began introducing speakers. First up were the candidates for class treasurer, in alphabetical order. Next were the aspiring secretaries. Then came the vice presidential candidates.

I couldn't pay attention to any of the speeches. All I could hear was Mr. DiNapoli announcing someone's name, some applause, repeat repeat repeat. I was worried. What if the cape and the pins got screwed up? What if Dan, who was controlling my music, screwed up? What if I got up there, drew a blank, and screwed up my speech? Much to my surprise, I did not have to go to the bathroom. I took this as a good sign.

Mr. DiNapoli stood at the podium and said, "Our next candidate for class vice president is Jason Mulgrew." Students started to give their customary applause as Dan, right on cue, pressed *play*. Jimi struck the G chord on his Strat two times before going to C and then to D and then back to C. The snare drum blasted twice and the rest of the band joined in. I stood up.

("WILD THANG . . .")

In my head, I shimmied and swayed up onto the stage, cool and confident as could be, my fur cape glorious and flowing and green, a sight to behold. In reality, I walked with the gait of one about to shit himself.

When I stepped onto the stage, I stopped for a moment, my back still turned to the crowd, showing them the pin-covered cape. I untied the cape, letting it fall off my shoulders as if I were letting a towel drop off my body before skinny-dipping in a lake. There were some "woots." This made me happy.

With Jimi still singing "Wild Thing," I got to the podium and waved to Dan, which was our predetermined signal for him to cut the music. But the music kept playing. I waved again, this time more vigorously. But the music kept playing. I was frantic now. Every second counted, and the extended music was cutting into my speech time. I thought about the warnings from Mr. DiNapoli about the consequences of straying from one's pre-approved speech. Thinking that Dan was either fucking with me or not seeing my waves (and hoping to Jesus Christ Himself it was the latter), I yelled into the microphone, "Dan, shut it off already!"

The music finally stopped and the audience laughed, thinking it was part of the act. With nowhere else to go, I rolled with it.

I wish I could tell you that my speech was awesome and inspiring, part Vince Lombardi and part Adolf Hitler (but without all the bad Hitler stuff). Instead, I answered imaginary attacks from the other candidates, who (I claimed) said that yes, we get it, Jason Mulgrew can give you 236 pounds of class vice president because he's got nothing else to do. But is he too fat to be an elected official, too chubby to serve on student government? Would he (let's just come out and say it) be too consumed with food to be your vice president?

I guffawed and said that this was not the case, that it would never be the case. That I was fully committed to student government and that nothing could distract me from that—not now and not after I was elected. These attacks were specious and baseless. And while I said all this, I pulled out a Tastykake from my jacket pocket and ate it slowly, savoring it. As I ate, crumbs falling from my mouth, I said again and again that these were lies, that I was ready, willing, and able to be on student government, that I'd be the best class vice president the school had ever seen. Or at least in the top twenty—definitely good enough for top twenty. Once I finished the Tastykake, I said, "Wait, what was I talking about again?" I got in one last "Mulgrew for VP!" and walked away from the podium. Dan restarted Jimi.

Though I thought it went well, there was no relief after the speech. In fact, I was more stressed out than ever, as students spent the rest of the day voting. There was nothing more I could do. The posters had been hung, the buttons had been distributed, the speeches had been given. I could only sit back and wait.

The votes were counted well into the evening in a closed-door meeting involving Mr. DiNapoli, the sitting student government officers, and a representative each from the junior, sophomore, and freshman classes, all of whom had been sworn to secrecy. By the next afternoon, the results were in. Before they were released to the entire school, Mr. DiNapoli met with each candidate individually to let them know if they had won or not, in the same order of the speeches: treasurer candidates first in alphabetical order, then candidates for secretary, and so forth.

Slowly, the results trickled out. By the time Mr. DiNapoli had finished meeting with all the treasurer candidates, we knew that Chris Simmons, the clear favorite going into the race, had won. Likewise, once the secretary meetings were wrapped up, we knew that Conor Pollack, while not the clear favorite but thought to have some chance, would be the next class secretary.

When I was called into Mr. DiNapoli's office, I knew that none of the VP candidates who met with him before me had been told they'd won. As an *M*, I was in the middle of the order in which the candidates would meet with him. This, coupled with the fact that there would be two vice presidents, should have left me feeling pretty good about my chances. Instead it was like sitting through a two-week-long trial and knowing the verdict was finally ready to come down. Not a trial for like murder or anything, but maybe trespassing.

"Jason, come in, sit down," Mr. DiNapoli said, motioning to the chair opposite his desk, the same one I had sat in a week before when I talked to him about Dan's crazy sign idea.

"As you know, there were a lot of terrific candidates in this year's election." (Uh-oh.) "This was especially true in the race for class vice president." (Crap.) "I've only been moderator for a few years now, but I can tell you that this was the closest race I've ever seen." (C'mon, c'mon, c'mon, c'mon.) "So I have to thank you for taking the time and making the commitment to run. It was a great campaign." (*Crap.*) "And I also have to congratulate you: you've been elected vice president of the senior class."

I was speechless—a rare occurrence. Mr. DiNapoli added, "One other thing. Look nice on Monday. We're taking a picture of the newly elected student government officers for the cover of next year's Prep brochure."

The Prep brochure, which had arrived in my mailbox almost five years earlier, the one I'd memorized front to back, back to front, the one with the student council officers looking like goddamn future senators on the cover.

Maybe it was time to get a haircut.

St. Joseph's Preparatory School

INFORMATION
BOOKLET

1996 – 1997

acknowledgments

The following people are incredible at being good humans:

Michael Signorelli, Erin Malone, Rakesh Saytal, Brendan Caffrey, Jerome at the Hotel San Regis, Selena Strader, those friends who read and provided comments on various incarnations of this book, and my family.

All at once: thank you and I'm sorry and God bless you and let's have a drink already.

About the author

2   Meet Jason Mulgrew

3   Six Questions for Jason Mulgrew

About the book

10   Questions to Prove that You Jerks
Have Actually Read My Book:
236 Pounds of Class Vice President
Edition

Read on

15   An Excerpt from *Everything Is Wrong with Me*

Insights,
Interviews
& More . . .

Meet Jason Mulgrew

JASON MULGREW was born in Philadelphia. He moved to New York City after graduating from college and started a blog in early 2004, when most people were still using AOL.com e-mail addresses.

His first book, *Everything Is Wrong with Me*, was published by Harper Perennial in 2010 to much critical and commercial success, having been purchased in more than fourteen states in two time zones. Once, someone took a copy on a flight to Paris. (Pretty fancy, huh?)

Nicole Goddard

He lives in Brooklyn with his wife and enjoys traveling (to first-world countries only), watching sports while drinking beer, and standing nude in front of the mirror. You can find him online at www.jasonmulgrew .com. ∿

Six Questions with Jason Mulgrew

Why did you write 236 Pounds of Class Vice President?

I wrote my first book, *Everything Is Wrong with Me*, in large part to get back at all the people who either wronged me or doubted me, as well as all the people that lived above the buildings that I was hustlin' in front of who called the police on me when I was just trying to make some money to feed my daughter.*

But with one book—I mean, anybody can get lucky once, right? Somebody knows someone and things fall into place and boom—you've got yourself a book published. Not to mention that the Internet has changed everything; these days, everyone with a Tumblr blog or a Twitter feed with a minimum of one thousand followers is required by the Constitution of the United States to get a book deal. Seriously. Look it up.

(If you are reading this in the future—like, in 2015—Tumblr and Twitter were two types of social media. We went batshit crazy for them back in the day. Ask an older person to tell you about them.)

However, two books? That's a little harder. That means that either enough people bought your first book (note: I said "bought," not "enjoyed") to ▶

---

* Editor's note: This is a Notorious B.I.G. reference. Jason assumed that everyone would get this, but we made him let us put this in.

make it worth the publisher's while, or you had to swallow enough pride to take an 80 percent decrease between your first and second advances. I won't tell you what it was in my case, because that's not important.

So I guess I wrote *236* to prove to everyone that it was not all luck. No, it was 98 percent luck and 2 percent willingness to do whatever it takes to prove to everyone that it was not all luck.

Another reason that I wrote *236* was the more Hallmark-y reason: I had no idea what I was doing as a teenager. None. I was a weird, sensitive nerd. All my role models were tough men whose primary use for books was throwing and whose feelings ranged from "tired" to "angry" to "drunk" and back again, with "ball-busting" the common thread throughout. Now, I don't have any delusions about reaching out to overweight teen nerds everywhere and changing their lives, but if one kid will read this and say to him- or herself, "Hey, maybe one day I, too, can write a shitty memoir about this time in my life!," well, then, that's how you measure success in my book.

(Pun not intended, but embraced.)

What was the hardest part of writing the book?

The hardest part of writing the book was the loneliness and the self-doubt and the access to the Internet. Not in that order—maybe access to the Internet is first. You're sitting at your computer and you're on a deadline and you think to yourself, "Maybe I'll just check ESPN for the score of the Phillies game while these ideas percolate." Then four hours later, you've exhausted the Internet's supply of videos with the words "amateur," "college," and "threesome" in the title and are combing through Wikipedia, reading about the personal life of Anaïs Nin and the Isle of Man and Cnidaria, with a grand total of fifteen words under your belt. Then it's time for bed. I swear to God I would be on book number eighteen by now if I had a goddamn typewriter.

So the hardest thing about writing this book was the hardest thing about writing anything: writing it. Actually doing it. It's fun to talk about being a writer at parties, but what it comes down to is this: You want to be a writer? Well then, write. Put some goddamn words on the page and get after it already.

What was your most embarrassing memory to write down?

You know that saying, often used as a wedding toast, that starts, "Dance like there's nobody watching"? When you write a memoir, especially a memoir about something as potentially embarrassing as being a teenager, you have to write like there's nobody reading. It would be very ▶

Six Questions with Jason Mulgrew
(continued)

difficult for me to write about, for example, masturbating with my hand basically wrapped around my prostate if I thought about my mom reading that passage.

It wasn't a question of specific memories (although the pictures of me in the cape *did* make me cringe a little bit). It was rather that acceptance— getting to that place where you say to yourself, "You know what? Fuck it," and you whole-ass it. If you half-ass it or avoid the unpleasant moments, you're doing a disservice to yourself and your story.

Can you talk about some of the themes you were trying to explore?

I don't know if we can call what I did in this book "exploring themes," but obsession is a big part of the memoir. As a teenager, you are a jumble of nerves and hormones and insecurity. For me, and for many of my friends, this jumble manifested itself into various obsessions. I don't mean "obsessions" in a scary way—like stalking the hot local news anchor or whatnot—but getting heavily invested in different hobbies or interests.

Almost every chapter has an obsession as its theme: Brutus, masturbation, the Prep, music, girls, a successful student council run, etc. Though I (perhaps obviously) have no formal training in writing, I know that when you tell a story, you always start with the motivation or drive of the central character. But what I found interesting, as I thought about what stories to include in this book, was how varied— but equally strong—those desires were. Back then, whatever I wanted at that time was the single most important thing in my life. Now, the single most important thing in my life is probably the half Xanax I take every night before bed.

(And my wife, of course.)

Do you have any advice for aspiring writers or for high school students who have read the book?

Did you combine two questions into one in order to stick to the six-questions limit?

Yes.

Well done.

Thank you.

So there's another question after this one?

Yes.

Got it.
Advice . . . for aspiring writers, read a lot. And write a lot. And develop a very thick skin.
For high school students who have ▶

read the book, that's tougher. I guess if you identified with some of the things that I went through, to borrow a phrase, it gets better. Probably. Whatever high school is for you—the best time, the worst time, the most boring time, the most I-can't-wait-to-get-away-from-my-parents time—remember, it's only four years. Well, it's only four years unless you flunk a grade. Or skip a grade. But you know what I mean.

However, I can't answer this question and not cover the following: If you are an attractive high school girl, pardon my bluntness, but please go to the nearest nerd and fuck him. Or at the very least make out with him.

See, the odds are that the jock you are dating has peaked. He will never be as cool and successful as he is right now. I can promise you this: his two-touchdown performance in the homecoming game isn't going to matter very much when you're thirty or forty.

On the other hand, I'm not saying every nerd at school is going to grow up to be a handsome millionaire. But— and this may be hard to believe—it is the nerds who rule the world. Because things like intelligence and creativity and curiosity, which may or may not manifest themselves in awkward behaviors or strange hobbies during one's high school years, are major assets in real life.

So go out there and fuck or make

out with a nerd. Just think of the
karma! I am neither handsome nor
a millionaire, but if any of the girls
I lusted over in high school had slept
with me back then, I would still be
sending them monthly checks today.
Not huge checks—again, I'm not Bill
Gates over here—but maybe $50
a month, or even $100, if I had a
particularly good month gambling.
Just a small regular "thank you" for
MAKING MY LIFE as a sixteen- or
seventeen-year-old.

Only my advice here. Take it or leave
it. And no matter what, practice safe sex.
Obviously.

As the title is **236 Pounds of Class Vice
President,** *we have to ask: what are you
weighing in at now?*

Fuck you. ～

Questions to Prove that You Jerks Have Actually Read My Book:
236 Pounds of Class Vice President Edition

by Jason Mulgrew
Author, Raconteur,
and Man-About-Town

IN SUPPORT OF MY FIRST BOOK, I made a promise: if your book club read my book and held its meeting in an accessible location in New York City, I, personally, the author himself, me, would attend that meeting and discuss the book with your group.

Some people called me crazy. "What if," they asked, "you showed up and it was all a ruse, and you were robbed or stabbed or worse?" So be it, I said. An artist must suffer for his art. And anyway, if any shit went down, I can shriek like a banshee. Loud. Very loud.

I had limited experience with book clubs, and did not know what to expect. Though not concerned with my personal safety, I *was* worried about the level of awkwardness that might come with sitting in a stranger's home, talking about a book that (at best) only half the people in the room had read. So in order to facilitate conversation, I devised a list of discussion questions, which I'd pass around at the start of the meeting. If nothing else, perhaps it would expose those freeloaders who didn't read the book and showed up only for the dip.

Though the discussion questions were for the most part successful and inspired some conversation, I learned quickly that there isn't much book discussion that goes on in book clubs. Sure, for the first fifteen or twenty minutes people might ask me about the book and we'd go over the questions. But once the alcohol kicked in, I'd find myself sitting silently in the corner, plowing through bean dip and drinking wine, listening to book club members (mostly women, with a number of gay men here and there) talk about weird sexual things their significant others tried to pull on them after they came home from a night out drinking.

So, obviously, it was a blast, and a worthwhile experience. I plan to make the same crazy promise for *236 Pounds of Class Vice President*. But this time, I present to you the discussion questions in advance. Good luck.

1. What is the tenth word on page 89? No looking, please.
2. What kind of dog was D'Ogee?
3. What combination masturbation technique was not possible, because of too many moving parts?
4. Starting on page 51, I describe Mr. Kearney, the disciplinarian at the Prep, as follows:

   If Miss Piggy were a man—a really, really pissed-off man—and one bad motherfucker, she'd be Mr. Kearney. Country strong, with heavily pomaded hair and thick glasses that gave his eyeballs a bulging appearance, ▶

> *Mr. Kearney told us how things were going to go. With his face alternating between the various shades of red and pink you'd find in a Hallmark store in early February, he said we were to adhere to the dress code (blazer, shirt, tie, slacks), we were to be on time, we were to be respectful, and we were to be Christ-like. As he said each of these things, he pounded his cantaloupe-size fist on the lectern, showing off a class ring with a purple stone larger than either of my balls. With his multiple Christ references, he struck me as some sort of goddamn Catholic vengeance warrior; I could see him in a fit of rage suddenly breaking a chair over a student and screaming "May Christ have mercy on your soul!" in Latin.*

     Pretty good, right?

5. Why is it that I was the firstborn, but my younger brother Dennis was named after my father? Please be brutally honest in your assessment.

6. Creamed chipped beef. Not a question, but I just wanted to remind you guys that it's terrific.

7. What is the thirteenth word on page 89? Again, no looking.

8. After being introduced to and falling in love with the Beatles, I wrote that my appearance started to change, and I started to look like John Lennon—if he had let himself

go and moved into a _____ for four years. Please fill in the blank.

9. What was your favorite song in high school? If you lost your virginity to a song, what song was it? While we're on the subject, please discuss how you lost your virginity, regardless of whether or not a song was involved. If you could describe your deflowering while wearing this Ronald McDonald wig, I would really, really appreciate it.

10. God, I love wine. You guys look great, by the way.

11. What the fuck was my dad thinking when he got me a motorcycle? I mean, *seriously*? It was such a blatant and misguided attempt to "man me up" that he might as well have come home and given me a handgun or a tiger for my sixteenth birthday.

12. When was the last time you had a Kahlúa and Cream? They really are quite good.

13. What was the name of the Mouseketeer with whom I was briefly in love?

14. How weird is it going to be the next time I see Alison? She's, like, a real person, with whom I still keep in touch.

15. But c'mon—when you were in high school, would you have considered dating a guy who wore a cape?

16. And why did I wear a cape in high school, anyway? (Note: There is ▶

no real answer to this question.
I would just like your opinion/help.)

17. Why was it that I chose vice president over the other student council positions?

18. Do you guys have any Pepto? This wine is giving me some killer heartburn. Are my teeth purple? ∾

An Excerpt from *Everything Is Wrong With Me*

Chapter One: A Break, A Beginning

It was the summer of 1973, a great time to be young, dumb, and in my father's case, full of Budweiser, Quaaludes, and reheated pizza. That lost generation—born too late to be hippies, too early to be disco freaks—strutted up and down the streets of my parents' South Philadelphia neighborhood, a grid of row-home–filled streets filled with working-class Irish Catholics and some Polish Catholics, bounded on the south by the Walt Whitman Bridge, the sports stadiums, and the Navy Yard; on the east by the mighty Delaware River; on the north by fancy Society Hill and, farther north, Center City; and on the west by the worst border of all: the Italian neighborhood that, thanks to Rocky, South Philly would become famous for in a few years. Sporting impeccable Afros and now-ridiculous but then-cool hairstyles—the men looking like Rod Stewart or Eric Clapton and the women like "Crazy on You"–era Ann or Nancy Wilson, but without all the trappings of fame and talent and good looks—and in their hip clothes, members of that tween generation joined friends hanging out on the corner, drinking beers, and listening to Bad Company, Derek & the Dominoes, and Mott the Hoople. After getting done with work, there wasn't much to do aside from getting drunk ▶

and listening to music. Which was fine for just about everybody involved.

My dad, Dennis Mulgrew, had just graduated from St. John Neumann High School, on Twenty-sixth and Moore streets. He was tall and lean, slowly beginning to collect tattoos, and was without his trademark mustache that he would wear throughout my lifetime. He wasn't my dad at the time—he would be "blessed" with his firstborn six years later, one year after marrying my mom—but rather just some guy who liked to drink, chase women, listen to rock 'n' roll, work on cars, and look good. In short, your typical teenager, fresh out of high school, not quite ready to embark on adulthood, instead occupied with more pressing and immediate matters, all in some capacity relating to narcotics and/or pussy.

He had recently gotten a job on the waterfront in Philadelphia, where he and pretty much every guy he knew worked as a longshoreman,[*] but on the weekends during the summer my dad would head "down the shore" to North Wildwood, a small island off the Jersey Shore, exit six on the Garden State Parkway, where his entire South Philadelphia neighborhood transplanted itself every year from Memorial Day to Labor Day.[†] There, in this quaint beach town filled both with large Victorians and kitschy and colorful motels, united by a miles-long boardwalk dotted with fudge and taffy shops, pizza parlors, and of course, all the carnival games and rides, he shared a shore house with a dozen or so other guys from the neighborhood, guys with names like Franny, Billy, Frankie, and Mikey and nicknames like Shits, Tooth, Flip, and Porky. Neighborhood guys, solid guys, genuine guys; guys who had known each other since kindergarten, guys whose fathers had all grown up together, guys whose

---

[*] To this day, I'm not exactly sure what longshoremen do. I think it has something to do with taking cargo from ships that come into port on the Delaware River and putting half that cargo in warehouses and selling the other half to your friends on the cheap. Also, there is a lot of cursing, napping, drinking on the job, and complaining about your wife involved. I could be wrong, but I'm pretty sure that's the basic gist of it.

[†] Whether this is because North Wildwood had more bars per square mile than any other shore town in New Jersey is unknown, but presumed.

understanding of the world outside their neighborhood was limited to the names of things they were smoking (Acapulco Gold, black Afghani hash, Hawaiian indica, etc.).

Just hanging out by the fish tank in a three-piece suit, jacket off, about to pour a can of beer into a little glass. You know, normal, everyday stuff.

On this Saturday afternoon in July of '73, my dad and his friends, being good blue-collar young men of Irish Catholic descent, were taking part in the preferred activity of their fathers and their father's fathers and their father's father's fathers before them: getting messed up and doing stupid shit. This could take various forms, such as:

- getting drunk and starting fights (usually with each other)
- getting high and stealing cars for joyrides
- taking some pills and breaking into friends' houses to steal household appliances and throw them in the ocean or bay
- something involving poop (human or animal)
- all of the above ▶

Despite being a few months shy of his eighteenth birthday, my dad had made plans to spend that Saturday drinking with some friends from Third and Durfor, a corner hangout back in the city, at Moore's, a bar on the inlet that sat atop jagged rocks that jutted out into the Atlantic. But he was broke. The night before, he had loaned his buddy Charlie [pronounced CHA-lee] his last twenty dollars, which Charlie had promised to repay first thing Saturday [pronounced SAH-ur-dee] morning. But Charlie never showed up. So instead of going to Moore's, my father joined his friends in a much cheaper activity: jumping off the pier into the bay. That was the plan, at least. Never mind that they had been drinking (and probably doing other impairment-inducing things) since they had woken up. And never mind that the distance between the pier and the water below was not insignificant. And never mind that no one in their group had ever done this before. None of these facts was deemed a deterrent.

I'm not exactly sure about this, but I think that in the early '70s a man's manliness and testicular fortitude were symbolized by the pomposity of his hair. My dad was fortunate in this regard. The Mulgrew genes guaranteed that he and his four brothers sported the biggest and baddest white-boy Afros their side of Girard Avenue, huge auras of kinky hair that extended straight outward and upward, looking not like they had accidentally stuck their fingers in electric outlets and had been shocked, but rather like they intentionally stuck their fingers in sockets because they looked that. fucking. good. So when the group, now gathered on the dock, lingered there—looking over the water below them, tacitly waiting for someone to step forward and offer to make the first jump off the bulkhead into the green-blue deep below—his Afro firmly in place, possibly touching it up while Zeppelin's "Black Dog" blared on the radio, my dad happily volunteered.

Never much for oceanography (or really any graphy, except perhaps pornography), my dad didn't realize that as he was preparing to make his jump the waters of the bay were receding with the tide. He was aware of the existence of tides in general (probably), but at the moment he was more concerned with turning up the rock 'n' roll and "Boy, do Shelly's tits look great in that bikini" than the moon's gravitational effect on our oceans. Therefore, it probably didn't cross his mind, as he was taking off his shirt and pulling one last swill of beer, that the water below might not be very deep, possibly not deep enough to accommodate a diver, possibly not deep enough to accommodate a diver as tall as him. In fact, at five o'clock on that Saturday afternoon, the water was only about four feet deep. My dad is and was then six feet, two inches. Four feet of water, a six-foot-two-inch human being. That math doesn't exactly work out.

But then again, my father didn't know much about math, either, except that it was for nerds. Without a second thought, and to much less fanfare than he had hoped (what, no cheers? no hoots?), he stepped onto the bulkhead and dove off the pier into the shallow water below. I like to think that he looked like an angel as he fell,* descending gracefully toward the rich, dark waters of the bay, each ripple glistening in the sunlight on a beautiful summer afternoon. However, I realize it was probably closer to a gangly drunk seventeen-year-old awkwardly falling headfirst off a pier.

And then THUD. A small splash and then THUD.

Inebriated as he was, as he first hit the water my dad somehow had the presence of mind to use his arms and forearms to partially break his fall. Before his head broke the plane of the water and lodged itself into the muck at the bottom of the bay, his arms hit first, lessening the blow on his head and neck. I don't know if this was a conscious decision on his part or a primal reaction to that horrible overpowering feeling of dread that arises at a moment of crisis when the voice in your head screams "Do something, you asshole!" but either way, it saved him. After that initial splash, his body knifed through the surface of the water and his arms, forearms, and head planted in the bottom of the bay, like a boot stuck in mud; for a brief moment his legs stuck out of the water ramrod straight like a totem pole. Once the force of the impact had subsided and gravity began to take its toll, the muscles in his legs and lower back gave way, and his body crumpled and splashed lamely and limply into the water. ▶

---

* Albeit an angel with a juvenile criminal record.

The next thing my dad remembers is standing under the outdoor shower of his shore house, several minutes after the dive, washing the black bay mud out of his ears, his hair, and his clothes. He had no recollection of coming out of the water, climbing the pier to rejoin his friends, words spoken among them, or walking back to his house. But he was no stranger to the occasional blackout and, standing under the shower, everything appeared to be okay: he could see, he could feel his hands and his legs, and he still had his dick and his balls. With this much, life could go on.

Once back in the house, the afternoon wore on and my father kept drinking on into the evening with his buddies, despite a nagging pain in his neck. As the evening progressed, after dinner was served (the usual: several boxes of spaghetti and two jars of Ragú, split nine ways), the pain also grew. This was something new; usually the more narcotics he consumed, the less pain he felt. After all, that was the whole point of drugs and booze, wasn't it? Not only that, this was a new kind of pain. It wasn't a throbbing, it wasn't a burning, it wasn't a bruise, it wasn't an acute sensitivity. It was a deep pressure that started in the base of his neck and spread slowly to his head, shoulders, torso, and arms. The more he moved, the more it hurt, so he made a makeshift neck brace out of a sweatshirt, hoping that it would both provide support and restrict the mobility of his neck. But despite his ingenuity and the solid C+ he had received in biology his junior year of high school, his neck brace didn't alleviate his pain. Worse, despite his drinking, the pain got so bad that eventually he had to retire to his "room" for the night, which was not a room per se but rather a low-traffic area of the upstairs hallway, as all the beds had already been claimed for the night.

When he woke up the next morning, the pain was unbearable. His neck and shoulders had swollen and it was nearly impossible for him to move, talk, or even breathe. Realizing that his home remedy of sweatshirt neck brace and dozen beers hadn't been the panacea it had been in the past, he reluctantly decided that he needed medical attention. Yet there was a small problem. He couldn't drive to get this medical attention. Not because he didn't know how, and not because of the injury, but because his license had been suspended due to a run-in with the law when he was fifteen, two years earlier.* After his father beat him within an inch of his life for that debacle, he knew to stay away from driving. Drinking, drugs, and fighting were fine, but no driving. No sir.

Surveying the passed-out bodies strewn about the house around him in the early morning, with their bearlike snores and their hobolike breath, he knew that no one was going to drive him to the hospital. Calling 911 was out of the question entirely, because, as my dad would tell me over and over again through the years, "911 is for pussies." After mulling it over for more than six seconds, he "borrowed" his buddy Paulie's car and was shortly zipping up the Garden State Parkway, heading back to Philadelphia. It's not like he was joyriding here, he reasoned—this was a good excuse. And besides, it helped take his mind off his neck. Instead of focusing on the pain, he fretted about whether his father would find out about him driving illegally and punch him in the head. Several times. Hard.

When he finally arrived in Philadelphia, almost two hours after he left North Wildwood, neither his mother nor his father was at home. He called his aunt's house around the corner, assuming his mother, Anna (the former Ms. Anna Bodalski), would be there. When she picked up the phone he said, "Mom, I think I broke my neck" and explained what happened. My grandmother, arguably the most rational woman ever put on the planet and easily the most intelligent person to share DNA with me, could do nothing but laugh. Not because she didn't care or was unkind, but a broken neck? It wasn't possible. She assured him that Dennis, you did not break your neck. If it were broken, you'd know it and you wouldn't be talking on the phone, let alone sleeping and driving. Being the mother of ten children, she was used to assuaging worries and calming fears, so she promised him that she'd be home shortly.

When she got home a few minutes later and saw her son sitting upright on the couch smoking a cigarette, she, a Polack ▶

---

* My father would not get his first legal license until he was twenty-nine, despite driving a truck part-time for four years in his twenties. Don't ask, because I don't know.

so logical she almost single-handedly eradicated the Polish joke in America, began to cry. His neck, which had been swollen for hours now, looked like a water balloon ready to burst. It was badly bruised, with a purplish hue that extended from his neck up to his hairline and down over his shoulders. They got in the car in short order and were at St. Agnes Hospital within minutes.

At this point in the telling of the story, my dad is at the height of his glory. This is where he'll slow it up a bit for dramatic effect. He'll lean forward in his chair, take a long drag from his cigarette (a Marlboro Red, which he's been smoking two packs a day of since he was twelve), and tell you how when he and his mother got to St. Agnes, the doctor immediately took X-rays but, upon examining them, ordered another set to be taken. "Because the doctor," he'll continue, his Celtic cross glimmering, hanging just above the paunch that protrudes from his getting-ever-tighter white wifebeater T-shirt, "thought that the X-rays were a mistake." In his professional opinion, no person with such extensive damage to the vertebrae of his neck could be moving around, talking, and functioning like my dad was. Then my father will lean back in his chair, the now-faded tattoos on his forearms and biceps loosening with his recline, and continue on about how the doctors at St. Agnes rushed him to Hahnemann Hospital, at the time the finest in Philly, because of the severity of his injury. When he got to Hahnemann, the doctor there didn't believe the story about how he had broken his neck, how he had jumped into the shallow bay but kept drinking, then slept, then drove, then came to the hospital. So the doctor asked my grandparents about it. When they confirmed his story, the doctor, shocked, told them that in his twenty years of practice, he'd never heard of anything like it.* Amazing, he said. Then, turning to the anxious parents, "Mr. and Mrs. Mulgrew, it is a miracle that your son is not paralyzed. There is no other reason. It is a miracle."

With the story wrapping up, my dad will say, "So that's how I broke my neck and that's why I got this scar," pointing to a

---

* When I first started hearing this story, it was twenty years of practice. Soon it became twenty-five. Recently, I've heard it as high as thirty-five. By the time my children hear this story, the doctor will have been 110 years old with eighty years of experience under his belt and possibly there will be a shaman involved.

six-inch scar that runs from the base of his hairline down to just above the middle of his shoulder blades. If I'm in the room, or my younger brother, Dennis, or my little sister, Megan, is, he'll point us out and add, "And that's why you're gonna be rich some day," explaining to all those present that in order to "meld" (his word) the bones of his neck together, the doctors used three ounces of platinum wire, which is still in his neck, and which he has made abundantly clear numerous times over the years we can remove and sell to a jeweler upon his death. So I've got that going for me. Which is nice.

Right about now, any reasonable listener would expect a moral to the story. Perhaps something like "Don't jump headfirst in shallow water when you're drunk" or at least "Be sure to measure the bay before you get bombed and dive into it." But my dad spins it a different way, concluding, "And you know what? To this day, I never got that twenty dollars back from Charlie Edwards. If he hadn't borrowed that money, or at least given it back to me on time, I would have been down at Moore's drinking with the guys from Third and Durfor. I wouldn't have been sitting at home and never would have broke my goddamn neck. And he still hasn't given me that damn money. Christ."

[smokes cigarette, watches television]

"Son of a bitch."

[shakes head, smokes cigarette, watches television] ▶

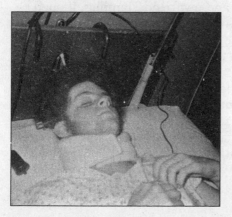

Not the best way to spend a summer, especially with twenty less dollars in your pocket.

Read on

Stories like this one are the kind of stories I grew up with.
Many of them started with "I remember one time when we
found this box of horse tranquilizers . . ." and ended with
"And that's when I learned that it's good to know Spanish in
jail." Unlike a lot of people my age, I never heard about my dad's
high school football glory days and his big interception in the
Catholic League championship game. I didn't learn about how
my Uncle Joey won the science fair in eighth grade with his
project about the moons of Jupiter. My mom never told me about
how she and my dad met at the local ice cream parlor and over a
root beer float fell madly in love. I didn't get the stories about how
my grandfather worked hard at the mill after the war to support
his growing family.

Because none of this happened in my family. My dad did play
high school football, but he was more interested in booze and
petty crimes than the nickel defense. My Uncle Joey never won
any science fair, but he did get arrested on Thanksgiving—twice
(I'm not sure what that has to do with science, but it's pretty
impressive nonetheless). The first time my mom laid eyes on
my dad, he had just been stabbed and was bloodied but was too
drunk to care or really even notice. And my grandfather, God rest
his soul, was officially a small-time grunt running numbers for
Philly's Irish mafia and unofficially one the greatest entrepreneurs
in the whole neighborhood.

Growing up, I thought this was all normal. I didn't know any
better (hey—I was just a kid) and my friends' families, though
maybe not quite as colorful as mine, certainly had their fair share
of characters and stories. That was just the way it was. It wasn't
until high school that I began to realize that my situation was
unique. Because I was a nerd,* I got a scholarship to a private high
school outside the neighborhood. It drew students from all over
the Philadelphia area—students whose parents were pharmacists,
lawyers, teachers, and bankers; who lived in houses with lawns
and swimming pools; whose families didn't steal cable and who
had never seen their father fistfight another man at a sporting
event or on a random Tuesday. Hell, their parents didn't even say

* Still am.

things like "motherfucker" and "prick" and "This shit is for real" in front of them. Strange, but true.

But life was never boring because we always had stories. And really, isn't that what it's all about in the end—the story? the memory? the ridiculous experience that you lived through, that you rehash to hungry audiences at parties and in bars and in holding cells? Stories that make everyone around you gape in delight, howl in amazement, buy you drinks, and yell for more? I think so, and I'm sure my dad does as well. And I hope you do, too, especially if you just shelled out money for this book. Because otherwise, you're totally beat for that cash. So let's just try to make the best of it, okay? ❧

Don't miss the next book by your favorite author. Sign up now for AuthorTracker by visiting www.AuthorTracker.com.